PLANET X

THE 2017 ARRIVAL

BY

DAVID MEADE

Published by eBookIt.com

ISBN-13: 978-1-4566-2692-1

ii

TABLE OF CONTENTS

FROM THE AUTHOR

This is the most important book you will ever read in your lifetime. It contains information which can save your life, and those of your loved ones.

Information is power. That is the reason why governments often withhold information from the public.

Revelation of an event is often progressive. You need to take this book and use it as a guide to do further research on your own. Don't stop learning – your life may depend on it.

I hope you enjoy reading this as much as I enjoyed researching and developing it. Research is my hobby and I've written three Investigative Books – one on Enron, one on the economic debacle of 2008 and this one. There is no comparison – this one dwarfs the others in terms of its importance.

My first book was published by Aspatore Press, a division of Thomson-Reuters. My second book was self-published, both in print and as an eBook. Because of time considerations and the importance of this subject matter, I first published this volume as an eBook. The mysterious death of Dr. Robert Harrington has been addressed as well.

Godspeed and Smooth Sailing!

THE IMPACT THREAT OF NIBIRU – PLANET X

The existence of Planet X is beyond any reasonable doubt, to a moral certainty. In this section we'll examine proofs of its existence. In fact, if you want to ask one simple question that posits the theory of the reality of Planet X, just ask yourself where did 2.2 Trillion disappear to in the Pentagon's budget, and why do we have over 100 Underground Deep Bunkers throughout the U.S.? Jesse Ventura discovered this. Why are critical government infrastructures moving from their susceptible positions on the East Coast to the protected areas of Colorado?

But let's look at the astronomical evidence. I have seen Planet X on the Wide-field Infrared Survey Explorer (WISE) through WorldWide Telescope. This is a NASA infrared-wavelength astronomical space telescope, launched in December 2009. It is currently in the constellation Pisces, and is clearly marked as an Unidentified Object (but quite plainly visible dark red star) known as IC 5385.

If you'd like to view it yourself, you can install WorldWide Telescope. Just Google it and you'll be right at the page. It's an observatory on your desktop and the most sophisticated online program I've seen. You can view in multi-wavelength views and see stars and planets in context to each other. You can take a guided tour of the universe. Microsoft Research has posted this amazing program online at no cost in memory of Jim Gray. Jim was a researcher like myself. He was manager of Microsoft Research's eScience Group. Jim Gray has gone missing, probably as a result of a sailboat accident.

We're not going to see Planet X or Nibiru, by whichever name you prefer to call it, until the month it reaches perihelion with our Sun. The exact week and month of its arrival is detailed within this book. Time is now very short.

Planet X is estimated to orbit our Sun with a period of slightly less than 4,000 years. Keep in mind it cannot be seen with normal telescopes. You need to be in the infrared spectrum to see it now, as it approaches south of the ecliptic. On Microsoft WorldWide Telescope go to the constellation Pisces and switch view mode to WISE. You cannot miss it. It's highly inclined with an eccentric orbit to the ecliptic plane (the plane around which our planets circle the sun). It is red and dense like Mercury and somewhat larger than Saturn.

Science Digest has reported that this red dwarf star was sought by the Pioneer space probes and there are national newspaper reports that it may have been found by the Infrared Astronomical Survey Satellite (IRAS) as soon as the early 1980s. The government has known about it since then. All major governments – Russia, China, England, South Africa and of course ours as well as major European heads of state are totally aware of it. Of course Scriptural references to it in Matthew 24, Revelation 6 and 8 all point to damage estimates of major impacts.

John Moore's national radio show has had real-time Intelligence on Planet X. He's interviewed (discreetly) ex-Navy veterans who have moved to the Ozarks on the recommendation of their commanders. They were told what is going to happen but not the details. What you're reading here is the details and the deep background. If you haven't got Windows 7 or 8 and thus cannot install the astronomy program, just watch this video and it will take you through the story and you'll see it for yourself. [1]

https://www.youtube.com/watch?v=NectwKUxN5g

THE EMP EFFECT AND PLANET X

CLASSIFICATIONS OF SOLAR FLARES

The highest impact solar flares normally experienced are X-Class flares. M-Class flares have a tenth the energy and C-Class flares have a tenth of the X-ray flux seen in M-Class flares. The more powerful M and X class flares are often associated with a variety of effects on the near-Earth space environment. Y-Class exist, however they are off the charts. In this section we'll discuss how Planet X will create these flares and the effects on the Earth.

HOLE IN THE EARTH'S MAGNETOSPHERE

You probably know about the hole in the ozone, but do you know about the hole in the Magnetosphere? Governor Jesse Ventura did an excellent show on this.

The area around the Earth that extends beyond the atmosphere is called the Magnetosphere. The Magnetosphere begins at approximately 1,000 km and extends thousands of kilometers into space. The sun and the Earth's magnetic field create this layer.

NASA has discovered (THEMIS mission) that the earth's magnetic field contains a hole, which is ten times larger than they previously believed. The magnetosphere shelters us from solar flares, but the hole is now four times the size of the earth.

RISK ASSESSMENT

The worst case scenario would happen when a violent CME that accompanies a Y-Class solar flare would both come at us in a 1-2 sequence. A Y-Class Solar Flare would send the part of the earth exposed to it back to the 1850s. It would have an EMP effect on all circuitry and electronics. Only hardened military electronics would survive.

EFFECTS OF Y-CLASS SOLAR FLARES

An EMP, or Electro Magnetic Pulse, is generated from the detonation of a nuclear device and also by extreme solar activity, such as that which was experienced in the year 1859. In the late summer of 1859, a great solar storm hit the planet. This storm was the product of a coronal mass ejection from the Sun.

On September 1st and 2nd, 1859, the Earth's inhabitants experienced the greatest solar storm in recorded history. It was called the Carrington event. This storm short circuited telegraph wires and caused massive fires. The typical light show you can watch in the far north, known as the Aurora Borealis, was seen as far south as Cuba, the Bahamas and Hawaii. Spark discharges even set the telegraph paper on fire. The electrical grid at that time was in its infancy, consisting mainly of a few telegraph wires in larger cities. This event, though frightening to those who witnessed it, had no major impact on the society of that day.

Planet X's arrival will create an event that occurs on the surface of the Sun that releases a tremendous amount of energy in the form of a solar flare or a coronal mass ejection, which is an explosive burst of very hot, electrified gas that has a mass that exceeds that of Mount Everest.

This time events will be different. The event will bring down the electrical grid, and the shelves on the grocery stores will be cleaned out inside of a day. Banks and ATMs don't work without electric current. Gas pumps won't be functioning. Food transportation will stop. Rioting and looting will be unrestrained. Communications satellites will be down. The 911 function on your phone isn't going to work. For as long as it lasts, until new transformers can be built or imported, society will be in chaos. This will be the calling card of Planet X upon its near approach to our Earth and Sun. [2]

THE NEBRA SKY DISC AND PLANET X

A 12-inch bronze disc, the Nebra Sky Disc, may be the key to determining the last recorded time in history that Planet X flew by the earth. The Disc is bronze and inlaid with gold. The Sky Disc of Nebra was found near Europe's oldest observatory in Goseck.

The story of the Disc is an unconventional discovery. It was found in 1999 by German treasure hunters using metal detectors inside an ancient forest. However, German law dictates that all such relics are state property. Instead of turning it in, they attempted to sell it on the international market in 2002. Working for the Swiss government, archaeologist Harald Meller posed as a straw buyer. The Disc was then seized by authorities.

THE OLDEST KNOWN IMAGE OF THE COSMOS
It contains symbols including the sun, the moon and the constellations Pleiades, Capricorn, Pereus and Gemini. The Disc reflects the sun in the position of a solar eclipse. What is causing the eclipse? It's not the moon, which is on the opposite side of where it should be located. The best deductive logic is that an unknown near-earth object is causing the eclipse.

The 32-centimeter Disc weights approximately 2 kilograms and is decorated with gold leaf symbols reflecting the four planets – Venus, Mars, Jupiter and Mercury. Using modern-day astronomical software we can plot the actual day and hour the Disc was produced. The result is April 6, 1810 BCE at 8:30 AM. The Disc was thus created during the Bronze Age, 3,826 years ago.

There is a curve bar at the bottom of the river bank – the Saale River, which still currently exists. Much of the Disc's current coloration is green due to a tarnishing effect. The discovery site is a prehistoric enclosure in the Ziegelroda Forest, known by the name of Mittelberg or "central hill," located 60 kilometers west of Leipzig. The treasure hunters stated it was found within a pit in the forest near the ancient Goseck observatory.

A very interesting point discovered upon examination is that the Disc reflects the constellation Orion at the very bottom. Orion should not have been observable from that location in Germany at that time. The nearest location from which it could have been observed is Luxor, Egypt, further adding to the mystery. The only logical conclusion is that the 26 to 30 degree difference is accounted for by a simultaneous pole shift.

The mystery-shrouded Sky Disc of Nebra is thus an advanced astronomical clock. It's a compendium of knowledge from the earliest times. It gives us the best proof of the existence and passage of Planet X in the ancient world. The purpose of the Sky Disc is no longer a matter of speculation – it is the oldest visual representation of the sky in existence, and it postulates the approach of a near-earth object, causing the daytime eclipse, as well as causing a major pole shift that has just occurred – 3,826 years ago. [3]

THE PLANET X POLE SHIFT

The Magnetic poles have been shifting towards Russia at the rate of over 40 miles per year. But when Planet X arrives the physical poles will shift, creating a variety of tsunamis, hyper-earthquakes of never-before experienced magnitude, and the shifting of land masses.

In this section we'll examine the Sun States of California, Arizona and Florida and how they will be affected. We'll look at Colorado and we'll also examine one area of the U.S. which is prime for relocation and survival.

California is composed of rock plates. When rock is compressed, it breaks, and you'll see the crumbling of rock as its mass is pushed upwards. The West Coast will have major catastrophic changes which will combine to synergistic proportions. Tidal waves will assault the coastline. High hurricane-force winds will erode the coast. Earthquakes will occur on the fault lines. Volcanic activity will erupt over pools of magma. Forest fires caused by volcanic explosions and severe lightning storms will spread throughout the region.

California is highly dangerous. The primary long-term danger in riding out the shifts will be for those who are living near the boundaries of plates. Cities without mountains as a backdrop will be washed out to sea. Very high-level quakes will create broken gas and water lines. Chaos will ensue. San Francisco is situated on the San Andreas and related fault lines. A lot of the population will be trapped. Water will overflow the city. Survival will be minimal.

In Arizona and its vicinity, the Hoover Dam will likely not survive the high Richter levels of earthquake activity. Rivers will flood and the areas within twenty miles of rivers should be avoided. Much of Arizona in cities such as Phoenix will be unaffected with the exception of wind damage or volcanic dust. Arizona has withstood many a pole shift.

The area around Tucson is safe. It is surrounded by mountains. Rainwater just runs back into the lowlands. Flash floods may create lakes, but Arizona is well beyond any danger from tidal waves.

Florida's highest elevation is approximately 105 meters in the Florida Panhandle. This part of the state may be able to survive. As for central and south Florida and the Florida Keys, these parts will be devastated by a tide of water.

Colorado is a very safe area. The domestic division of the CIA is relocating to Colorado. That should give you many clues right there. Eastern Colorado descends into plains and has various rivers from the hills onto these plains. Some of these streams and rivers will be a port of safety in the aftermath of a pole shift.

The Ozarks (northern Arkansas and southern Missouri) have an average elevation of 800 to 1,000 feet. They are self-sufficient with water supply and have ample agricultural land. They will do well. They are also isolated from large metropolitan areas, and absolutely well enough inland to avoid flooding through tidal waves. [4]

THE VATICAN PLANET X OBSERVATORY

The Vatican is in a very strange business. It owns a Large Binocular Telescope (LBT) in Arizona on Mt. Graham, near Tucson. It is a Near-infrared telescope. The cover story is that they are looking for extra-solar planets and "advanced alien intelligence." It is reported that the instrument is chilled to -213 Celsius, or -351 F, to allow it to conduct near-infrared observations. Planet X is of course a dwarf star, a binary twin to our sun, and can best be observed in this spectrum.

Why all of this interest by the Vatican? What are they really interested in? It is reported by a Father Malachi (a Vatican insider) that he knew about an inbound planet which would cause the destruction of millions of people. He said it would look like a red cross in the sky when it appeared. He had indicated it would appear between 2015 and 2025. The exact date is a secret, hidden in the archives of the Vatican.

Father Malachi was Professor of Paleontology at the Pontifical Biblical Institute of the Vatican. He was known to have criticized the Vatican for not releasing the full content of the "Third Secret of Fatima." The presumption has become that the Vatican didn't want the facts known about Wormwood.

In an interview with Art Bell on national radio, Father Martin was asked why the Vatican was so deeply involved in the exploration of "deep space." He responded, "Because the mentality amongst those who are the highest levels of Vatican administration and geopolitics know...what's going on in space, what is approaching us..."

Another Vatican insider stated that the Graham Observatory is used to study "anomalous celestial bodies approaching the earth." He compared it to what the CIA did with one of its secret eyes, also known as the twin to Hubble. It is called SkyHole 12 or KeyHole 12.

He stated that during an Alaskan radio telescope monitoring of a deep space probe (a program called SILOE), that a photograph had been taken of a huge planet moving closer inbound to our system. This was way back in 1995. He was told the information was classified well beyond "Top Secret." He further stated that the probe was created in Area 51 with an electromagnetic pulse motor, and that it was put in orbit by a space plane. Its purpose was to approach Nibiru or Planet X and transmit information back to the radio telescope.

The fact is that the Vatican has a Secret Intelligence Agency named the S.I.V. and this stands for Servizio Informazioni del Vaticano (the Information Service of the Vatican). The fact is that the Vatican's telescope and its twin are mounted at the focus points of the LBT's two giant 27.6-foot-diameter telescope mirrors. Each instrument is cooled to -351 Fahrenheit in order to observe in the near-infrared wavelength range. What do you think they are looking for? [5]

PHOTOGRAPHIC ANALYSIS OF PLANET X

In this section I'd like to discuss photographic analysis of a couple of videos and still photographs. The photograph is that of one taken by a Glenn Vaughan, on April 11, 2015. It was taken at sunset while flying down the Pacific side of Central America.

Professor Vaughan produced a 20 megapixel RAW file with a Sony digital camera. He confirmed the report by sending it in for analysis. He took it through an aircraft window. An analysis of the metadata indicated it was taken at 240 dpi (high resolution), and the metadata further show the date, time and GPS coordinates. A Sony DSC-HX50V camera was used.

Using an analysis in Starry Night (Professional Version), the analyst did a field of view analysis. See the video "PLANET X SYSTEM UPDATE FOR NOVEMBER 2015" (https://www.youtube.com/watch?v=A0va3_v0zi8). He determined Mercury is above (superior conjunction) with the sun. The object in the field of view is not Mercury. Mercury does not have "wings" as does this object, and Mercury does not have the brightness or contrast of the unknown object.

The analyst also performed a Gamma Test on a paint program. Gamma testing shows if an object is hot or cold. Hot objects generate their own light (like our sun). A lens flare would be a cold object, for example. They redirect light, but do not generate it.

He started with a Gamma of 1, reducing it by steps. At .10, the Gamma illustrates stark differences. At .01 all we have left are the hottest, brightest objects. He then reduced brightness to a value of -3, and contrast to +2. The hot objects fade in together, proving they are genuine.

Then he ran a luminance test to determine the intensity of the object, and he determined it was behind the sun (not a reflection). There are wings on the object, which is clearly behind the Sun.

This is, as he stated, "deeply disturbing." The reason is that gas tails of comets follow the solar winds. They are not visible until the comet's orbit is inside of Jupiter, at which point the luminosity of the Sun makes the tails of the comet (or planet) visible.

The object (which is clearly Planet X by deductive logic and process of elimination) is therefore now between Mars and Jupiter! It is not in the "outer reaches" of the solar system. It can clearly be seen, especially from the southern hemisphere. As of April 2015 Nibiru is now close enough to the Sun that the Sun's light is reflected in the tail, making it appearing like wings.

MELISSA HUFFMAN'S UNKNOWN OBJECT VIDEO ANALYSIS
Location: Sanibel Causeway Florida – Melissa Huffman, photographer

The object is interesting – she pans the sky and shows the object and then the moon. It's clearly a planet visible just north of the Sun.

Using Stellarium, and knowing the known moon phases (comparing it to her video of the moon), you can go to the derived date, and take away the atmosphere and ocean, and you can determine the relationship of the stars to the horizon.

Let's first determine the date. If you follow the moon on charts it is well below the horizon in all dates after September 26th. Therefore the video was taken earlier in September. September 2015 moon phases give us a comparable to the moon in the video. If you look at the moon phases diagram to the actual picture of the moon, it has to have been photographed on September 23rd based on the phasing of the moon.

On that date Mercury in Stellarium is very dim, somewhat above the horizon, and in a different location than the unknown object. All other planets are below the horizon – the video therefore plainly shows an unidentified planet of some magnitude quite clearly. I can reach no other conclusion other than this is a rare photo from this hemisphere of Planet X approaching us. [6]

WILL THE REAL PLANET X PLEASE STAND UP?

Science admits quite a few points in favor of Planet X. Our solar system is surrounded by a vast collection of icy bodies called the Oort Cloud, located 50 to 200 thousand AU (Astronomical Units, or the Earth-Sun distance) from our Sun. If our Sun were part of a binary system in which two gravitationally-bound stars orbit a common center of mass, this interaction could disturb the Oort Cloud on a periodic basis, sending comets towards us.

Science admits an asteroid impact is absolutely responsible for the extinction of the dinosaurs 65 million years ago. And science admits a comet may have been the cause of the Tunguska event in Russia in 1908. That explosion had one thousand times the power of the atomic bomb dropped on Hiroshima, and it flattened 80 million trees throughout an 830 square mile area.

Science admits a recently-discovered dwarf planet, named Sedna, has an extra-long and usual elliptical orbit around the Sun. It has an orbit ranging between 76 and 975 AU (where 1 AU is the distance between the Earth and the Sun). Sedna's orbit is estimated to last 12 thousand years. Sedna's discoverer, Mike Brown of Caltech, has stated that Sedna's location doesn't make any sense.

"Sedna shouldn't be there," said Brown. "There's no way to put Sedna where it is. It never comes close enough to be affected by the Sun, but it never goes far enough away from the Sun to be affected by other stars." What does this mean? It means a massive unseen object is responsible for Sedna's mysterious orbit.

Now Brown is part of a Caltech team that claims to have discovered the ninth Planet in 2016 (Pluto has been demoted in the last decade and is no longer considered a planet). Brown's calculations suggest Nemesis or the Ninth Planet is 10 times the mass of Earth.

Science knows that binary star systems are common in the galaxy. It is estimated that one-third of the stars in the Milky Way are binary systems. Red dwarfs are also common – in fact, they are the most common type of star in the galaxy. The existence of Planet X is therefore quite in line with the current knowledge base.

But let's wrap it up – the so-called Ninth Planet recently disclosed in January of this year by the NASA people is a combination of information and disinformation. It's what we would call in the Intelligence Community a bush league type of propaganda. My advice – don't major on minors. Don't study their announcement or pay too much attention to it.

The real Planet X is in the vicinity of Mars and fast approaching Earth. It has no direct relationship to this recent "discovery" and this so-called discovery is just one more piece of evidence that the powers that be are running scared, and making mistakes as they go. [7]

THE REVELATION SIGN OF PLANET X IN 2017

Insider reports dating back to the 1980s indicate that the Planet X system will dramatically impact the earth with meteor showers, seismic activity and a pole shift. Insiders have stated since this time that the arrival of Planet X would be in the second half of the second decade of the 21st century – 2016 to 2020.

The primary reason we have not been able to discern the very time of its arrival is that we have separated science and the Bible. The two actually mesh perfectly.

Here is a critical piece of information on its timing:

"A great sign appeared in heaven: a woman clothed with the sun, with the moon under her feet and a crown of twelve stars on her head. She was pregnant and cried out in pain as she was about to give birth." Rev.12:1-2.

The great sign of The Woman as described in Revelation 12:1-2 forms and lasts for only a few hours. According to computer generated astronomical models, this sign has never before occurred in human history. It will occur once on September 23, 2017. It will never occur again. When it occurs, it places the Earth immediately before the time of the Sixth Seal of Revelation.

During this time frame on September 23, 2017, the moon appears under the feet of the Constellation Virgo. The Sun appears to precisely clothe Virgo. Only the one that occurs in 2017 constitutes a 'birthing.' This is when planets traverse within the legs of Virgo. Jupiter is birthed on 9-9-17. The twelve stars at that date include the 9 stars of Leo, and the 3 planetary alignments of Mercury, Venus and Mars which combine to make a count of 12 stars on the head of Virgo. Thus the constellations Virgo, Leo and Serpens-Ophiuchus represent a unique once-in-a-century sign exactly as depicted in the 12th chapter of Revelation. *This is our time marker.*

The next event that follows is that Planet X will fully eclipse the Sun and cover the whole earth and full moon in shadow on the next full moon date. This is Thursday, October 5, 2017 (the following full moon date). Our clue that this is the date is from Revelation, Chapter 6, verse 12:

I watched as he opened the sixth seal. There was a great earthquake. The sun turned black like sackcloth made of goat hair, the whole moon turned blood red... Rev. 6:12.

The sign of the Red Dragon (Planet X) appears in heaven immediately after the sign of "the woman." Planet X is visible during this total solar eclipse, as the following verse indicates:

Then another sign appeared in heaven: an enormous red dragon with seven heads and ten horns and seven crowns on its heads. Rev. 12:3.

The Book of Revelation is a cryptogram set by the Almighty, and it is solvable, according to the Book of Daniel at the very end of time. The mystery of the timing of the arrival of Planet X is "hidden in plain sight" and right before us. [8]

REFERENCES

Footnotes 1 to 8 – Meade, D. (2016, January 29). Planet X News | News about Planet X / Nibiru and its possible effects on our weather and world – Planet X News. Retrieved from http://planetxnews.com/

For your reference this is my website:

http://writers-web-services.com/px

And my email is DavidMeade7777@gmail.com

There's some fascinating photographic analysis on my site – I encourage you to visit it.

A DATE OF DESTINY

Could it be that government insiders from NASA, DoD, and the CIA know that 2/3 of the population of the planet could perish during the coming pole-shift caused by the passage of Planet X?

Is it possible that every secretive government agency in the USA is fully aware of what is expected? The Vatican is fully abreast of what is happening. The public is not being warned and has virtually no chance to prepare.

Is money to be held in higher regard than people?

Expect increasing levels of disinformation from the mass media.

Planet X has high mass and a high level of magnetism. It disrupts the surface of any planet it passes. 7 years prior to its passage, its electromagnetic influence changes the core of the earth, and triggers major weather changes, seismic activity and volcanic eruptions. Unexplained sinkholes and bizarre weather accompany its return. So called "global warming" is a cover story. The earth is heating up from the core outward because of the approach of Planet X.

During the 20th century, there were about 35 volcano eruptions in a typical year, and now there are more than 35 eruptions every single day!

Only if you understand the truth can you hope to understand the incomprehensible events taking place today.

Like many of the stars in our universe, our sun is a member of a binary system. Planet X is a brown dwarf star that is on an elliptical orbit every 3,600 years around the sun.

The amazing thing about this book is that I believe I have discovered not only the year but the month Planet X wrecks havoc on the world. It is so close you will be shocked when you realize it. The world as we have known it will no longer exist.

"All truth passes through three stages. First, it is ridiculed. Second, it is violently opposed. Third, it is accepted as being self-evident." – Arthur Schopenhauer

ISAIAH 24:1 KING JAMES VERSION (KJV)

24 Behold, the LORD maketh the earth empty, and maketh it waste, and turneth it upside down, and scattereth abroad the inhabitants thereof.

How is it turned upside down?

I believe the next North Pole will be in Siberia, northwest of Lake Baikal, near the borders of Russia and Mongolia. There is a large magnetic anomaly there. The famous Tunguska blast event of 1908 occurred there. The Magnetic North Pole is already moving in that direction. Founders of the pole shift theory feel this is the likely location of our next North Pole.

If you want absolute proof of the existence of Planet X, aka Nibiru or The Destroyer – here it is – from the Book of Revelation, Chapter 6.

12 And I beheld when he had opened the sixth seal, and, lo, there was a great earthquake; and the sun became black as sackcloth of hair, and the moon became as blood;

13 And the stars of heaven fell unto the earth, even as a fig tree casteth her untimely figs, when she is shaken of a mighty wind.

14 And the heaven departed as a scroll when it is rolled together; and every mountain and island were moved out of their places.

15 And the kings of the earth, and the great men, and the rich men, and the chief captains, and the mighty men, and every bondman, and every free man, hid themselves in the dens and in the rocks of the mountains.

Think on these matters – a great earthquake, the sun darkened, the moon changed to the color of blood, and stars falling. The atmosphere splits open while every mountain is moved. That's when the underground bunkers prepared for the elite are operational.

Only a Planet X passage could split open the atmosphere making the sky "depart as a scroll" and create these phenomena via its physical interactions with Earth. It could cause a worldwide earthquake moving every mountain and island via a pole shift. Therefore to me the description of Seal 6 in the Bible is the single greatest proof of Planet X. The timing of its appearance is immediately in sequence with the rapture. To understand this concept, let's go over some Biblical principles. Each chapter carries you further into the truth, and the final chapter, The End Game, ties together the entire cryptogram of Revelation in a way that indicates it can only have been designed by God.

A GENERATION

What constitutes a generation? Jesus plainly stated that "this generation shall not pass until all of these things are fulfilled." Is a generation 120 years, or 100 years, or 70 years or 50 years? One year fulfills all of these requirements.

2017 fulfills all of these time spans. 2017 less 120 years is the first Zionist Congress of 1897.

2017 less 100 years is 1917, the Balfour Declaration for Israel.

2017 less 70 years is 1947, when the UN General Assembly voted to create Israel.

2017 less 50 years is 1967, when Jerusalem was recaptured.

Is God awesome or what? No matter what length of years you choose, the end result is **2017**.

God answers the question of a generation in the Word, and all answers agree with each other. This is incomprehensible.

Beyond that, it's in the Scriptures. What date did Noah enter the ark? The 17th day of the 2nd month. 2_17 or 2017. Why is this in the Bible unless God wanted to disclose it? Jesus plainly said it would be "as in the days of Noah." It's in the Bible for a reason, hidden in plain sight.

God doesn't lie and cannot lie. Time is up. 2017 is a true Jubilee, which in Greek is "Yovel." This means "to be carried off."

I believe I may know the month. There is no restriction on knowing the month. There is certainly no restriction whatsoever on knowing the year.

Surely the Lord GOD will do nothing, but he revealeth his secret unto his servants the prophets. – Amos 3:7

Paul makes the distinction between the Church (whom he calls "ye", which is the plural of "you") and the unbelieving world that has rejected God ("they"). This distinction is critical to understanding Bible prophecy as Jesus Christ employs the same language in the prophecies in the Gospel. Paul reassures the church in Thessalonica that yourselves, meaning the Church, *"know perfectly"* that the Day of The Lord will come unexpectedly. At that point *"they"*, meaning the unbelievers of the world, will suffer *"sudden destruction"* and **"they shall not escape."** Paul then says in the very next verse **"But ye, brethren, are not in darkness, that that day should overtake you as a thief." The day is not a surprise at all to believers who are actively watching.**

REVELATION 3:3 NEW INTERNATIONAL VERSION (NIV)
Remember, therefore, what you have received and heard; hold it fast, and repent. But if you do not wake up, I will come like a thief, and you will not know at what time I will come to you.

Just turn this verse around, and it says if you are holding fast and awake, YOU WILL KNOW THE TIMING OF THE LORD'S RETURN AT THE RAPTURE. Not down to the minute, but you can know at a minimum the month and the year. That's scriptural. Revelation is progressive.

The rapture happens when the stars are falling in Rev 6:13 "And the stars of heaven fell unto the earth." These stars are made up of 2 things that are headed to earth – nuclear missiles and space rocks or meteors.

Jonathan Kleck heard from the Lord that we go up when the missiles comes down. Then in Rev 6:14 "And the heaven departed as a scroll when it is rolled together;" this is the nukes exploding and the mushroom clouds spreading the sky like a scroll.

I know a man who had a vision in 1977 that at the end of October of 2017 Russia and China would cause nuclear terror in a first-strike attack on the U.S. and Britain. In the vision, the attack occurs between midnight and dawn and it occurs by the very end of October 2017. It celebrates (to them) the 100th anniversary of the Communist Revolution.

Pope John Paul II suggested in the past that World Communism – the major powers of which are Russia and China – would make some hostile moves if they knew the secret of the incoming Planet X. Father Malachi Martin agreed and wrote in his book, The Keys of This Blood: "were the leaders of the Leninist Party-State to know these words, they would in all probability decide to undertake certain territorial and militaristic moves against which the West could have few if any means of resisting."

An invasion of Central Europe, South Asia, and the Middle East would probably occur at this time by Russia and China. This may be one reason the Russians are preparing to invade Ukraine.

WORMWOOD STAR

Authors Note: This is a most unusual vision I came across which substantiates the science we're covering, and the Bible itself:

Greetings in the Name of Jesus Christ, our Lord and Savior whom we long for His glorious appearing... All along I have heard and read about the Wormwood star in Revelation... Revelation 8:10-11 – And the third angel sounded, and there fell a great star from heaven, burning as it were a lamp, and it fell upon the third part of the rivers, and upon the fountains of waters; And the name of the star is called Wormwood: and the third part of the waters became wormwood; and many men died of the waters, because they were made bitter.

Wormwood Vision... However, as I was seeking the Lord through prayer and fasting through this past May (2005), The Lord showed me a powerful vision. I was taken to the heavens above the earth and I could see the earth in a distance not so far but clearly. Suddenly I heard a very loud sound and vibration with a very heavy zoom sound coming my way, hundred times louder than that of a Boeing 747 Jet. I then saw a very huge rock almost the size of the moon zooming past me at a very high speed with a large tail of fire besides and behind it. In the vision I was made to feel the waves of vibrations and the heat it generated that hit me, but I was protected. It was like someone was holding me by my collar and snatched me out of its way to a safe distance. I then could see other splintering rocks falling of this huge burning object and catching fire themselves.

I looked where it was headed. I knew it was headed for planet earth and thought well, I hope it lands in the ocean, not many people will die. However, it seemed targeted to the ocean and the United States. I was like, no Lord no! However, it kept its course targeting the ocean and the United States. It seemed as though someone was controlling its path unhindered and sure to hit its target. The vision ended. I shared the vision with a Prayer Warrior

Sister who then told me I had a vision of Rev 8:10-11. Wormwood Vision... However, I simply wrote the vision down and well, like we do with many visions, I kept this one and did not give much attention to it until this morning. This time I had a series of visions early in the morning as I got up to pray and all related... I was shown the Calendar... First, I saw someone dressed in a white robe holding a calendar and said to me, "The date has been set back for Wormwood" which I understood as forward...he then showed me the calendar with September/**October** written on it and the number 7. I did not see the year but I instantly knew it was the Fall Season. Then I was shown the second vision. People were having their usual life, and for some reason I saw people going to get movies at Blockbuster and did not seem to care about what was coming. People were busy watching movies. No one was warning anyone, life was going on as normal. Then in the same vision I was taken to the Chiefs Football Arrowhead Stadium, here in Kansas City. I knew it was fall and the beginning of the Football Season. In the Vision it looked like evening and the stadium was parked to capacity with everyone putting on their red Chiefs outfits, the stadium was all red as it is normally here with Chiefs games in Kansas City.

The game was going on as usual and suddenly there was a very huge loud bang in the sky...and then a huge object I had seen previously in my May vision lit the sky with red fire and zoomed passed across the sky...with very power vibrations that threw everyone off their seats and shook the stadium...

Everyone in the stadium begun to scream and go hysterical, and run to and fro, but could not leave the stadium. The huge ball of fire flew from the east across to the west. I knew it was Wormwood. The Stadium officials seemed to have had a clue about the coming disaster but did not warn anyone and also they locked the stadium doors and no one could go out. I then saw something amazing – people begun to fall on their knees and pray to God, they knew they were going to die anytime soon. I even saw young toddlers who had come to the game with their parents praying too. I was then taken back to the Blockbuster place and people saw the object and heard the bang and vibrations but seemed not to care about it, and some said, " I will die watching a movie"...

After this, I was shown the debris and damage floating all the way on the Atlantic ocean like the force of the moving star caused the debris of destroyed buildings to float all the way from Europe and dumped it at the east coast of America with more excessive damage...I remembered the Tsunami, it looked like very small compared to the Wormwood damage...

Angels High Tech Room...

I was then taken to a room were I saw men around some very high tech gadgets with screens that looked more like flat screen computer panels...but this was very high tech, technology I have never seen before...I knew the men were angels, they were all dressed in white robes and they turned on one flat screen panel and showed me the trajectory of the Wormwood star, then I was shocked to look and behind Wormwood was another star following the same trajectory, just a short distance behind Wormwood and again headed for the Ocean and the United States...they all seemed to have the United States in their path...

Uganda/Africa... After this I was then taken to Africa, in Uganda where I come from. I was shown people, very sad, looking to the destruction in the West. In Uganda, the staple food is Bananas (Matoke in Luganda, a Ugandan Language); people grow different types of Bananas and Plantains. However, the Path of Wormwood dried up all the banana plantations. There was already a drought before Wormwood and now the little food was gone. People cried because the West that provided them food was in destruction... fiercer hunger loomed on the horizon... However, I saw the saints gathered for Jesus Christ too in numbers...I was encouraging the saints there... I did not see as much destruction from Wormwood in Africa as I saw in the West in the Visions, apart from a fierce hunger and famine across the continent...

Source:

http://christian-forum.net/index.php?showtopic=18850

MISSILES OF OCTOBER

Larry W. Taylor posted on September 7, 2002:

Thought I would relate a dream to you I had in the early 1980's. I had been in prayer for about a week and on the last day of my prayer fast, I became very sleepy. So sleepy, that I couldn't keep my head up. I struggled against sleep as I was trying to hear the Lord. I lay back on the bed for a minute, but as soon as my head hit the bed I was asleep. I immediately had a dream. DREAM: In the dream I saw a long line of missile silos. I had the knowledge in the dream that these were missiles no one ever thought would be fired! I could not tell who the missiles belonged to. As the dream progressed, I saw the missiles begin to fire and lift out of their silos. I noted that not all the missiles were fired, but only a certain number were fired. I saw the missiles fired, go up into the sky and reach a peak; then start to arch and fall back towards the earth, towards it's target. I did not see who or what the target was but as the missiles began to fall towards their targets, I heard in my dream what sounded like a loud audible voice,

"THESE ARE THE MISSILES OF OCTOBER!"

This voice woke me up fully awake and I sat up with a start! I still remember in vivid detail the dream of the Missiles and the voice that declared they were of October. (No year was given).

http://standeyo.com/NEWS/12_Prophecy/120917.missiles.of.Oct.html

PUTIN DEVELOPS SUBMARINE DRONE TO ATTACK U.S. HOMELAND

By David Poortinga On 12/26/15 at 10:58 AM

"Russia is developing a drone submarine that can deliver a nuclear warhead or dirty bomb (that targets electronics) at U.S. coastal areas. Russian officials confirmed the existence of the program – dubbed Kanyon by the Pentagon – after images of a document detailing plans for the sub were broadcast on Russian state-run television."

See the complete article at:

http://www.newsweek.com/putin-develops-submarine-drone-attack-us-homeland-408707

This can be used for a nuclear first strike including an Electromagnetic Pulse (EMP) attack on the United States.

DUMITRU DUDUMAN:
CHINA AND RUSSIA WILL ATTACK AMERICA

September 3, 2013 1:40 pm By James Bailey

In April 1996, prophetic Christian minister Dumitru Duduman received a vision of a coming attack on America coming from both China and Russia. Dumitru Duduman was a Romanian native who came to America and founded the Hand of Help Ministries. He received many visions and dreams during his life. He went home to be with the Lord in May 1997.

The vision shared below is just one of many that Dumitru Duduman received. A complete list of his prophecies are available on the Hand of Help website, including the transcript of each one. Also included there are the prophetic words received by his grandson, Michael Boldea.

I prayed, then went to bed. I was still awake, when suddenly I heard a trumpet sound. A voice cried out to me, "Stand!"

In my vision, I was in America. I walked out of my home, and began to look for the one who had spoken to me. As I looked, I saw three men dressed alike. Two of the men carried weapons. One of the armed men came to me. "I woke you to show you what is to come." He said. "Come with me."

I didn't know where I was being taken, but when we reached a certain place he said, "stop here!"

A pair of binoculars was handed to me, and I was told to look through them.

"Stand there, don't move, and look," he continued. "You will see what they are saying, and what they are preparing for America."

As I was looking, I saw a great light. A dark cloud appeared over it. I saw the president of Russia, a short, chubby man, who said he was the president of China, and two others. The last two also said where they were from, but I did not understand. However, I gathered they were part of Russian controlled territory. The men stepped out of the cloud.

The Russian president began to speak to the Chinese one. "I will give you the land with all the people, but you must free Taiwan of the Americans. Do not fear, we will attack them from behind."

A voice said to me, "Watch where the Russians penetrate America."

I saw these words being written: Alaska; Minnesota; Florida.

Then, the man spoke again, "When America goes to war with China, the Russians will strike without warning."

The other two presidents spoke, "We, too, will fight for you." Each had a place already planned as a point of attack.

All of them shook hands and hugged. Then they all signed a contract. One of them said, "We're sure that Korea and Cuba will be on our side, too. Without a doubt, together, we can destroy America."

The president of Russia began to speak insistently, "Why let ourselves be led by the Americans? Why not rule the world ourselves? They have to be kicked out of Europe, too! Then I could do as I please with Europe!"

The man standing beside me asked, "This is what you saw: they act as friends, and say they respect the treaties made together. But everything I've shown you is how it will REALLY happen. You must tell them what is being planned against America. Then, when it comes to pass, the people will remember the words the Lord has spoken."

"Who are you?" I asked.

"I am the protector of America. America's sin has reached God. He will allow this destruction, for He can no longer stand such wickedness. God however, still has people that worship Him with a clean heart as they do His work. He has prepared a heavenly army to save these people."

As I looked, a great army, well armed and dressed in white, appeared before me.

"Do you see that?" the man asked. "This army will go to battle to save My chosen ones. Then, the difference between the Godly and the ungodly will be evident."

The events Dumitru Duduman saw coming are disturbing, but it is encouraging that he saw a heavenly army will be sent to deliver those who worship God with a clean heart as they do His work. This is consistent with many other prophetic warnings regarding war coming to America.

AUTHOR: JAMES BAILEY

James Bailey is an author, business owner, husband and father of two children. His vision is to broadcast the good news of Jesus Christ through blog sites and other media outlets.

http://Z3news.com

MORE PROPHETIC WARNINGS

Many other people have seen the same events coming that Dumitru Duduman saw. Man of them have also seen revival coming to America in the midst of the destruction. Here are a few examples.

Pastor TD Hale saw a rain of fire spreading like a flood across America, but in the midst of the destruction he saw a great revival coming first to America and then to the whole world as millions of beams of light shot up out of the flood waters. See more in my blog post TD Hale's 4th Prophetic Dream: Rain of Fire Coming to America (http://z3news.com/w/td-hale-4th-prophetic-dream).

Mike Shreve saw a worldwide nuclear proliferation coming among the enemies of America during the term of President Obama. He saw this was the scheme of the enemy to stop the spread of the gospel, which was being largely funded by America. However, God showed him "another third great awakening that is coming to our nation and ultimately to our world. So God has a plan that will outwit the enemy." See more in my blog post Nuclear Proliferation and a Third Great Awakening Are Coming (http://z3news.com/w/nuclear-proliferation-and-a-third-great-awakening-are-coming).

David Taylor saw a coming Russian invasion of America that would bring revival to America. The following is an excerpt from my blog post: David Taylor: Russian Attack on America is Coming (http://z3news.com/w/david-taylor-russian-attack-america-coming).

"What Americans don't understand is they think this is about Osama bin Laden and the terrorists. No, they are getting distracted. The real monster behind this attack is Russia. I can say that boldly because I know they are planning to attack America. Russia is planning an attack against America. What is going to happen is when this war hits America it is going to bring revival to America."

"I cannot tell you when the Russian invasion of America will happen because the Lord has not given me a date on that. All I can say to America is that she needs to get ready. In the middle of this war the greatest revival is going to break out because whenever darkness hits the glory of the light arises."

America's first president, George Washington, saw a coming invasion of America that would be the greatest trial our country would ever face. In his vision, he saw "vast armies devastate the whole country and burn the villages, towns and cities." However, he saw the armies of heaven coming to strengthen us. The angels "joined the inhabitants of America, who I perceived were well-nigh over come, but who immediately taking courage again, closed up their broken ranks, and renewed battle." As a result, "the dark clouds rolled back, together with the armies it had brought, leaving the inhabitants of the land victorious." See his complete prophecy in my blog post: George Washington's Prophecy of the Coming Invasion of America (http://z3news.com/w/george-washingtons-prophecy-invasion-america).

Way back in the late 1970's Chuck Youngbrandt received very detailed revelation regarding the coming invasion of America. He was not shown the timing of the events, although that has not stopped people from trying to guess. Like Dumitru Duduman, he also saw the attack will come from Russia and China. His revelation is extremely detailed so it clears up many questions. For example, the reason why George Washington saw America will ultimately be victorious is because we never officially surrender even though our military will be defeated and our nation will be occupied by millions of Russian and Chinese troops. After the fighting stops, American guerrilla forces continue to strike enemy targets throughout a season of occupation, which lasts for seven years. In the midst of these difficulties, Americans come into agreement seeking the Lord to deliver us. At the end of the seven year occupation, the prayers of the saints combined with the atrocities of the occupiers moves the hand of God to drive them out of our land. That is what George Washington saw when he said, "the dark clouds rolled back." Then the remnant of surviving Americans, which are the Christians, restore the battered Union. So America will ultimately be victorious, but at a very high cost, which includes a seven year season of servitude and bondage to ungodly rulers.

Dumitru Duduman saw that God "will allow this destruction, for He can no longer stand such wickedness." He did not see the final outcome of the invasion, but he saw the same army of angels that George Washington saw, going to battle to save God's chosen ones, those "who worship God with a clean heart."

MY COMMENTS

Ancient Israel went into bondage when they turned away from God, and it sounds like America is headed there now. When ancient Israel repented, God restored them and their nation. The sooner America comes to the point of repentance the better. Until we fix our relationship with God our problems will keep getting worse. But if we repent of our sins and pray and seek the Lord, He will deliver us. This battle is in the spiritual realm, not in the natural realm. When America wages war in the heavenly realm, then the battle will turn and God will deliver us.

EMP EFFECTS OF NIBIRU

David Daughtrey (10 Feb 2009)

January 1, 2009

Arcadia, Florida

Hello My Friend John Tng:

The first part of this letter is a little testimonial of where I am coming from. In 1994, while in my worldly ways, the Lord Jesus and the Holy Spirit finally got tired of my way of life and took me to the wood shed and whipped me severely. I thought they would kill me. I felt they had torn my heart out. I'm talking about mental suffering and pain. I cried and grieved so bad my heart felt like it stopped beating several times. He took one of the dearest things from me, my beloved dog. I loved that dog with a passion. She was my constant companion. Every step I took she was with me. She even slept with me. The problem was, I worshipped my dog and she worshipped me (like the Bible says, no other gods before me).

After several months went by, I was still crying daily over her mysterious death. She was very healthy according to the vet. No reason for her death. For several months before her I had a strong feeling that someone or something was controlling my life. Weird things would happen, not normal, like things fall down not up. Whoever heard of 3 or 4 flat tires on your car for 4 days in a row. It just doesn't happen. I could reach into a barrel of golden bb's and pull out the only lead one, three times in a row. I am serious. I decided that this was of a supernatural control. Then a friend of mine said, you know, it sounds like God is trying to get your attention. I said, OK, I'll get me a Bible and read it and see if this is my supernatural controller because I can't keep living like this, it's killing me. So I read my King James, and the more I read the more I craved. About 4 books into the Bible, I cried and begged the Lord for forgiveness, and he did. God's Holy Spirit came into my body in May of 1994 and cleansed my heart and soul. As long as I live, I will never forget that feeling, it lasted right at 8 hours. I felt like I had swallowed a dozen tranquilizers. I did cry sorrowfully, and was ashamed of my past. I know everyone has their own story.

Two years went by, reading my Bible day and night, and listening to sermons on the radio. I read where Daniel prayed three times a day. I can pray six times and I did. I lived and breathed God's word. My family thought I had lost my mind, I was so obsessed. Every minute I was alone, I talked to Him, like He was standing there invisible, of course. I believed He was anyway, even though He said not a word aloud. I probably would have croaked if He actually spoke to me like He did Moses. But I knew I was just a lowly servant, not worth much, surely did not deserve any special treatment.

Well anyway, two years went by, praying day and night. At the end of each prayer, I'd lay in bed and think about Jesus and ask Him the same question every night before sleep. Lord Jesus, will I live to see the rapture? And what did you really look like? Then one night while on my knees praying, I just started, when out of nowhere, like someone flipped the TV on, a picture appeared before my eyes. Flames of fire, dancing all around me moving just like real fire. Well, I knew I hadn't done anything wrong, so this had to be Satan trying to interrupt my prayers. I said, go away Satan, you're not making me stop praying, but they did not leave, so I got scared and jumped into my bed. My wife said, my that was a short one. I said, if you had seen what I had just seen you would have made it a short one too. Well, I decided that would never happen again – wrong! The very next night, the same thing happened again, but this time I was not going to quit, and I didn't budge. Then something happened right in the middle of those flames, they opened up and revealed a road, a long road, it was like I could not go but one way, and if I got off this road, I'd get burned, or my life's road would be through flames of fire. And believe me, the last 13 years has not been a picnic.

There were a total of 15 visions, most were at night as I began to pray by my bed. I had no control. They started and stopped when they wanted. Each and every night I thought it would be the last one. There was not a sound or word spoken in these visions, just silent pictures. Well, after three visions ended, I thought maybe I had pleased the Lord for Him to give me these visions, maybe a little special. Wrong!

Within 6 months. I had to file bankruptcy, one year later, a massive heart attack, and another year later another heart attack. I don't want to receive any credit for anything. I can't handle any more mental or physical suffering. I don't want God to think I'm getting puffed up. I must get on to the rapture vision.

The Rapture Vision of 1996

The coming events or warnings to happen just before the rapture takes place.

Dearly beloved doves, and brothers and sisters in Jesus name. This is the vision given in 1996 concerning world wide events to happen from one to three weeks before the actual Rapture. This I told to many friends, and was often hurt at the disbelief, and cold shoulders of many Christian friends that I finally just put it on the back shelf, and let it sit for the last 13 years. I now feel that the Rapture is so close that it needs to be told, urgently. I agree with our Christian brother John Tng that there have been so many dreams and visions of the Rapture that probably no one will believe this one either, but that is okay. All I pray for is everyone to remember. Like our Bible says, the vision will prove itself with time and testing. By remembering these visions, it will or can save you much needless suffering. The Lord does not want to hurt his bride.

There is going to appear above the earth (in the sky) a strange object, it will be large and sphere shaped, like a ball. It will look like it's been built in sections, like a football with huge rivets at the seams. Many will call this a UFO (Author's Note: Planet X), I don't know. It will have the color of copper or bronze. It will be on every TV around the world. People will be shocked like the World Trade Center. People will be glued to their TV's but you don't. The minute you see this, run to the closest food store and get enough canned food and bottled water for about three weeks, because between one and 24 hours after this object is seen world wide there is going to be a massive impact or collision on our sun's surface. It is going to happen on our blind side, we can't see it coming. It's going to be a super size twin asteroid hitting the sun in a vital spot, releasing a major solar storm, knocking out all of earth's electricity, all over the world. Those of you who read this will save yourself and your family much misery.

Now, this is what's going to happen in the next 2 or 3 weeks, while the electricity is off. Within a few days the whole world will start to go crazy with hunger. The banks and ATM's can't work without electricity. No gas pumps for food transportation, no refrigeration, total darkness. The robbers, rapists, and murderers will see right away that no one can call the police for help, because the solar storm has burned up all communication satellites, cell phones and telephones. The law can't even call each other. It's going to be total breakdown in large cities. There will be gun shots and screams all night long. Millions will be behind locked doors, praying and begging for God's mercy, for help and protection. He will answer millions of prayers. Millions will ask for forgiveness, and the Lord and his saints will perform millions of miracles during these three weeks.

Then after about 3 weeks of this, the Rapture will happen. Although there are different time zones around the world, the Lord showed it will be night here in Florida, USA. In this vision, I was taken in the spirit out in the woods behind my house in Florida. I didn't hear any trumpets or words (come up hither) but I know there's going to be, because the Bible says so. As I was standing there in the spirit of course, in the middle of some trees, a huge blue beam of light came down like a big flashlight, about a thirty foot circle around me. The blue light was identical in color to a welder's arc light at night. It was so blinding I put my hands over my eyes to see if I could see where or what this light was coming from. Then I noticed in the distant night sky, north, south, east, and west of me, blue colored stars jetting off the ground spiraling upwards traveling fast, they were heading for the bright blue object that was shining that beam of light down on me. It could have been the Lord or an open door to heaven. I really could not tell, it was so bright and blinding. Anyway, these little blue stars were going up in clusters. Different numbers depending on the size of town around me. Then all of a sudden when they reached about 9:00 high they burst into 10 times their size. Then I realized it was those alive in Christ joining those dead in Christ. I was too far away to see any new regenerated bodies or white gowns, but I'm sure they were. Then suddenly I was again taken out of my body off to the side to look at myself standing in the light. That's when I saw my own flesh glossing white as lightning, all my flesh, head and arms, my clothes remained the same. Then instantly, I vanished and my clothes fell to the ground. From what I saw in this vision, I was the last to go because all the alive in Christ had joined the risen dead.

The next afternoon, I had been reading my Bible for about an hour and then stood up in my living room to take a break, when suddenly at arm's length from my face appeared a blue cloud, the same color as the beam in the Rapture vision. In the shape of something like a football on its side, about 3 ft long and 2 ft high – white, a brilliant white 21 in the middle of it. I never did understand the meaning of the 21, maybe it was the day of the Rapture (Author's Note: I believe it was the number of years to the Rapture, 1996 plus 21 or 2017). I base no idea, your guess is as good as mine. Well, that startled me, so I sat down, shocked again because this was in the middle of the day and I wasn't even praying. Well, my son was in the kitchen making some tea and he said we were out of sugar. I said OK, I'll go down the road and get some. I began to sit up when this small newborn baby appeared before my eyes. I'm starting to come unglued now, because it hadn't been 5 minutes since the blue cloud, when this happened. I told myself, O my Lord, am I losing my mind? What, O God, does this baby have to do with the Bible? I must be losing my mind. I really started crying, it was blowing my mind. Then I noticed this newborn baby had this horrible growth on the side of its head, almost as big as the baby's head. It was scary ugly, that's how bad. After about 20 seconds, the vision disappeared, the baby was gone. I told myself I've got to get out of here or I'm going to start crying again. Well, I left and went after the sugar, about 2 miles away. I walked into the store, the owner and myself were the only ones there. He was on the telephone talking to some friend of his. I was at the counter waiting to him to get off the phone, when suddenly a young woman came into the store. She walked right up beside me and spoke to my friend on the phone. She said, you haven't seen my brand new baby have you and he said no ma'am. I happened to look down at the baby and nearly fainted. There was the same baby I'd seen 10 minutes earlier. My legs went like rubber and I had to lean on the counter to keep myself from falling. This action made this young woman mad, I mean angry, my reaction insulted her greatly. She yelled at me and said, the doctor says it's just a birth mark, and it will shrink and go away. She then stormed out of the store. My friend said, did you see that thing on the baby's head? I said yes, I saw it 10 minutes before I came here. That was the last one, never again in 13 years. The only thing I can figure out is, He was showing me that every vision he had shown me was going to happen. My friends, all I ask you is to remember what you read here. It's not important to believe this, but I beg you to remember.

Your brother in Christ

Unknown Servant

Source:

http://www.fivedoves.com/letters/feb2009/davidd210.htm

THE WORLD POST-PLANET X

April 23, 2013 1:11 am By James Bailey

In 1980, Ken Peters had a long, detailed dream about the coming tribulation period. This article provides the transcript and video of his amazing testimony. {Author's Note: He describes the EMP Event of Planet X and what transpires afterwards} I took the liberty of moving a few of his comments to organize them by topic so the transcript is not exactly the same as the video.

Today, Ken Peters and his wife Tonya are Senior Elders at the The Gathering @ Corona, which is an Apostolic Prophetic Reformation church.

There have been rumors circulating that Ken Peters no longer believes his own dream. However, I received an email from him stating, "Dear brother please post for me that I stand behind the dream completely, never have recanted." The IP address for the email matched the area where his church is located. I also spoke with Stan Johnson, founder of The Prophecy Club, which is where Ken Peters shared this testimony. Stan told me to his knowledge Ken never shared this testimony publicly before or after this recording. Even the day he shared it Ken was not sure if he would be able to do it because the memories were so disturbing. Stan said Ken never told him he had recanted his testimony. He is just reluctant to talk about it, which is understandable. So the rumors were not true.

TRANSCRIPT

At the time I received this dream I was not even a believer in Jesus Christ. I was raised in the Catholic Church but had never personally invited the Lord Jesus to come into my heart to be my Lord. As a practicing Catholic, I had no knowledge of what the Bible said about the tribulation period or any of the events of the last days.

When the dream began, I heard what sounded like a loud car horn. Then I saw people coming up out of their graves all over the world. People were not coming up out of every cemetery plot, just some of them. Even in the same cemetery there were other plots with nobody resurrected from them.

These resurrections were very violent. It was like the earth was receiving a small explosion and breaking open. I literally saw dirt flying. This was happening all over the globe. Those who were resurrected were clothed in white robes. It looked like they were wearing choir robes. The light glimmered off of their clothing. Their clothes and their bodies appear brighter than the sun. Their clothes made the men look very masculine and the women look very feminine. They looked mature but they did not look old. Those who had lost their hair had all of their hair back again. Young people who were resurrected were still young but yet had maturity about them.

All those who came out of the graves just disappeared. I never saw them go up into the clouds. They just vanished.

I did not see one single living person changed into a new body. I did not see any changes coming to any living person.

MASS HYSTERIA

As soon as the resurrected people disappeared from the earth mass hysteria spread across all the people left on the earth. People had the appearance of absolute despair. There was pandemonium everywhere. There was mass chaos, lawlessness, and fear everywhere. I was able to see in many quadrants of the earth and this was not just happening in one nation but it was all over the globe. The hysteria brought perplexity to just about everyone. Everyone had a look of hopelessness on their face. Nobody seemed to be happy about living. Lawlessness and fear permeated society completely.

No one was isolated from the despair that was hitting the world. No one was hidden from it. It was engulfing the whole globe. I was able to see into different regions and different continents and everyone was experiencing this. It was almost as though the whole world had become like a third world nation, completely behind the times.

It was like every person on earth had just left their mothers funeral. That's how people appeared. They were very grieving and despondent.

TWO WEEK SHUT DOWN

At that point, all electronic devices including televisions, telephones, radios, and computers all over the world were shut down for two weeks. I don't know what caused them to shut down. The shutdown caused people everywhere to be alarmed. It was very disruptive to businesses everywhere.

After about two weeks electronic devices started working again. However, everything had changed. The content of what was being broadcasted was completely different. The message being broadcast was depicting a coming new world government and leadership. They announced that a man to lead the new world government would soon appear.

THE NEW WORLD LEADER

The new world leader then appeared on television. He spoke with great eloquence and charisma. He was soothing and promised answers to all current issues. He was smooth and extremely convincing. He was able to solve nearly all problems. He was a consummate communicator. He explained how this removal of people was God's judgment upon them.

When Adolf Hitler spoke to the masses he had a demonic charisma that would draw people into his message. But that was nothing compared to this new world leader.

Since I was not a born-again Christian at the time that I had this dream, when I heard this man speak he began to convince me. This man was rallying the globe. It was very frightening.

Almost immediately he began to communicate through large-screen televisions that were strategically placed everywhere people met. All televisions were broadcasting the same message. When I had this dream in 1980, big screen televisions were not found everywhere. Neither were 24-hour news channels.

This man's messages were about new times that had come upon us as human beings. He gave new directives for global peace. He talked about the need for giving up national citizenship for world citizenship. Even though these things disturbed me, I was also being pulled into it in my dream. I began to really think about relinquishing my citizenship and this alarmed me greatly. Even though the message pulled strongly upon me I somehow was not convinced about this new order.

I constantly heard the term new order and world order and new times but I never heard him use the term new world order. I don't know why.

At staggering rates people were buying right into this plan that this man was releasing through the airwaves. There was no resistance. No one was fighting it. No one was saying anything publicly against it.

The man that I saw on the television, the man that could do signs and wonders and fix all the problems, I will never forget his face. As long as I live I will never forget his face. His face was almost supernatural in appearance. He was almost too perfect. For lack of better terms, he was the most handsome man I had ever seen. This man had everything going together for him, everything. He had kind of a chiseled look about his face. And everything about his appearance was almost perfect. When he spoke, there was a very strange quality about him.

Many years later, I read a scripture about the Lord Jesus Christ written by Isaiah the prophet saying Jesus had no comeliness or features that we would desire to behold him. In other words, Jesus was not some handsome specimen of a male. He was an average, rugged, probably different looking person. He was not the kind of guy that would be voted most likely to succeed on the GQ charts. But this guy that I saw was. He fulfilled that perfectly. Isn't it amazing that the antichrist would be the antithesis to Jesus? That he would be just the opposite of Jesus Christ?

Although this man did not act necessarily prideful he was very brash. He still carried the ability and the charisma about him to levy people into his situations.

This new leader was not resisted on the implementation of any of his policies, not one. No one stood up to challenge him. No one in America started a revolution. There was no resistance whatsoever, not on a grassroots level, not on a national level, no one.

Television continually, almost daily, explained to us that if we would align ourselves with this new order we would be saved from all of life's troubles. This is what this man said nearly every time he came on television. The new order was said to have all the answers to our problems and the leadership necessary to bring the change, causing the world to finally become the envisioned globe of peace. This is what we heard over and over.

I began to go into a serious depression. I began to ask myself if this was the end of the world.

ELDERLY EVANGELIST

I ran into an elderly gentleman. He was the first person in the dream that actually appeared to be friendly. You look like maybe he had some hope or might know what was going on. So I stopped him and asked him a few questions. I asked him, "Do you know what is happening in the world?"

He told me that the end was coming upon us and that he had not prepared for the times of the Lord. At this statement, sadness filled this man's countenance. He went from being joyful to being very sad. He said to me that he had not been right with the Lord. Then he began immediately to tell me God's plan for man's salvation. He carefully reached around in his back pocket and looked around over both of his shoulders to see if somebody was watching him. He pulled out a little pocket book, a little Bible, and he began to flip the pages to Scriptures showing me different things in the word of God about my need for Jesus to be my savior. He told me that I had to ask Jesus to forgive me of my sinfulness and my sinful nature. He told me if I would do this I would be given eternal life and that God's power would lead me during this life. He told me God would give me a victorious life.

I said, "Well, that sounds pretty good." I was convinced and so I prayed. I accepted Christ into my heart in my dream. This is a very strange occurrence to be a Catholic person, not knowing how to find Christ by the born again experience through the repentance of sins and receiving this in a dream. What's even more interesting is when I woke up from the dream I could not comprehend it. It took two weeks after the dream for me to become a Christian.

So I prayed the prayer with him. He put his hands on me and prayed some different prayers. As soon as this happened joy began to fill me. I did just as he said. I asked Jesus to forgive me for my sinful ways and to fill my heart with his presence.

There was something unusual about this man because he had a small following with him. These were people that had accepted his message that he was telling them about Jesus Christ.

A very unusual thing was occurring at this time in the earth. Babies were being abandoned just about everywhere. Almost on every street corner were babies being abandoned left in their little baby seats or baby baskets. This was very strange because they were anywhere from infancy to 16 or 18 months old. I could tell there were not any babies over the age of two.

We begin to pick up children everywhere and we began taking care of these children. I kind of joined up with this group of people because they were the only ones that seem to have any peace at this time anywhere in the whole earth that I had experienced.

Some very unusual things were happening with this group of people. It was amazing to me how they could meet people's physical needs. They would always run into people that were in need and they would meet their needs and then somehow lead them to Christ. No I didn't know how to do any of this yet because I had just hooked up with them.

In the dream my wife had also become a Christian, a believer in Jesus Christ. We were both hooked up with this man, helping him out.

At this point my wife and I began to really connect with this man and his followers. Some other things that were very strange to me was that things just somehow seemed to work out for this man and this group of people in the most unusual ways.

During the dream I didn't know what was going on. I did not know that God got involved in the affairs of men. As I grew up religiously I did not see God in that aspect. But I saw very unusual events happening with this band of followers. Food would multiply. Very unusual things would happen. They would pray for people and people would be healed.

WORLDWIDE REVIVAL

These so-called Christians were coming to the old man and his team of people and they were explaining how they once had a relationship with Jesus but had become cold in their faith and fell away from interest in a life of holy passionate pursuit of God. For a short period of time people were coming to Jesus in total surrender. I was able again to see above the globe and what I saw was very unusual. I got to see certain regions of the earth where light rays were just coming out high into the atmosphere. It almost looked like those big searchlights with the flame on the inside of them except that these were very brilliant, almost supernatural in appearance. After I saw these shooting out from the globe in many different directions I was given the ability to go down into these regions and actually see firsthand what was happening. Let me tell you, it was the most exciting thing I have ever seen. It is the very thing that gives me the determination to continue doing what I'm doing right now. If it were not for this part of the vision I would not minister in America anymore. I would move away completely.

Here is what happened. I began to see 12 regions in the United States of America and all over the globe where these beams of light would just come out and begin to shine into the atmosphere. When I got down close, what I saw was mass revival hitting the earth. I did not see any Ken Peters. I did not see any big-name evangelist or prophets or apostles for famous television personalities, Not one. All I saw was every day normal children of God ministering in the power like Jesus described in the Bible with the disciples. This was happening on a wholesale basis, everywhere! People were praying for sick people and they would be healed instantly. They would pray for blind eyes and they would be opened. They would pray for dead people and they would be resurrected. They were praying for the lost to come in. What I saw was the greatest thing I have ever witnessed since I have been alive. Nothing I have ever witnessed on earth could compare to what I was allowed to see. This period of time lasted about three or four months, maybe six months max. Maybe that long. It was so incredible! Regions were totally won For Jesus Christ.

What I saw was so incredible that it was almost unbelievable. Jesus said in John the works that I do you will do and greater works you will do. I did not see any greater works but I saw a greater quantity. I didn't see anything greater than raising a dead person, but I did see a greater quantity. It was almost like everybody was like Jesus walking around doing these works. You did not have to have a pulpit to stand behind to do this in this part of the dream. As a matter of fact I never saw anyone standing behind a pulpit. I think they finally understood the purpose of the ministry is equipping and releasing you to go out and be God's superstars.

This outpouring lasted for a short period of time and literally in regions there was complete light and right next door, which would almost be like a city next to it, would be complete darkness. There began to be an agitation in the spirit realm that was incredible.

EARTHQUAKES AND FAMINES

While I was on my way to make a business transaction a very unusual thing happened. There was an earthquake while I was on my way to the bank. I was just entering the bank. Across the street from my bank there was a large seven-story building. This was a triangle looking building. It was all glass in its appearance. In this dream, an earthquake hit and began shaking this glass building. It fell over and killed about 200 people. This earthquake was massive. I know from what I saw with the globe shaking at this point that it was a worldwide earthquake.

The earthquake hit and there were multiple, millions of lives lost. The world was completely stunned. The devastation to property and loss of life was beyond comprehension. It could not be measured. Some regions were so destroyed that they never bothered to send rescue teams in. That's how devastated they were. This destruction was global. It reached the whole globe.

The earthquake caused a massive change in weather patterns. The normal weather patterns completely changed. The patterns for winter became summer and summer became winter. You might have a day of snow and a day of heat. The world was in total chaos in this weather patterns. Predicting the weather became totally impossible. It was just useless to try to forecast weather. Predictions did not work.

Some very unusual things began to happen almost immediately. Crops began to perish due to droughts. I was able to see all over the glow the most fertile areas, the most fertile farming areas. I live at the time of this dream in the most fertile farming area in the whole world, the San Joaquin Valley of California. These areas were totally destroyed with drought and famine. Places that were once fertile were now arid deserts. It was almost hard to comprehend what I was seeing. It was almost immediate. It was like somebody just took things and twisted the whole order.

The thing that was strange to me was that weather seemed to have its own mind. The earth being shaken from its axis manipulated the weather. I was above the earth and I saw it shaking. I saw the earth rocking around like it was a drunken person trying to walk. It was very frightening to me.

I can't tell you how hopeless or empty I felt after seeing these things happening. Many times I wished I could have just woken up and pretended these things were not really happening.

NEW LAW ENFORCEMENT

Right about at the time that this earthquake hit, very unusual things began happening with the laws. I began to see local municipalities and no longer were the police departments the enforcers of the laws. But military police driving very unusual looking vehicles that I now know are called Humvees.

The vehicles I saw were black and were on just about every corner of every main thoroughfare. The man standing in the back was wearing a blue helmet.

I did some checking and learned that in 1980 there were no blue helmets or blue ball caps worn by any military on earth.

There was a big radio antenna or some sort of a device in the back of the Humvee. On the other side in the back there was a flag. It looked like the guy standing in the back had some sort of big gun. I was able to look inside and on the inside there was what I know now to be a laptop type of computer sitting on the dash. It had a computer screen that looked much like the airplane that I rode in today. They could look into this computer and it gave him all sorts of information.

What was unusual was that they were fairly peaceful. They were not rude. They were not mean to people. They were not obnoxious. I did not see any looters or anybody getting shot or anything like that. They seemed to be peaceful.

One thing that was happening was you could not cross state lines at this time without papers. Current papers were required to cross state lines. That was very strange to me.

At the same time I saw streetlight stands with little oval shaped cameras on top of them. In this dream it was revealed to me that these cameras knew the whereabouts of everybody's vehicles.

The changes took place almost instantly and with complete ease. There was peaceful martial law. Military vehicles were everywhere. They knew everyone's whereabouts. I found out how they began to know this. I did not understand this at first until it was revealed to me.

All of the nations of the world became as one. There were no longer any sovereign individual nations. Continents were no longer divided into countries but were divided into regions.

As time progressed in the dream, I was given the ability by being in homes that television sets not only broadcast and transmitted programming but they now had the capacity to actually send signals about what you are doing in your living room. I was able to see the television sets were actually watching people in their homes, monitoring their movements, monitoring their conversations. I was shown in the dream that the television did not even need to be on. It just needed to be plugged in. Later, I found out that televisions made after 1992 can in fact watch you.

The awareness of God being on the global scene was nearly impossible to detect. The global order had no presence of God in it whatsoever. Evil began to pervade every aspect of society. Darkness was everywhere. There was a clear line between who was God's people and who was not. You could walk down the street and you would know instantly who was who. It was not like it is right now when we sometimes wonder who is saved and who is not. This was so evident. There was a clear line of delineation. Spiritual demarcation was clearly seen.

PERSECUTION

At this point when all these miracles begin to happen this world order became very angry because what was happening was beyond their control. They were not able to manipulate it or stop it from happening. This makes the devil very mad. He gets very mad when we start functioning in the real power of the living God. That's when he really starts pulling all the stops out to try to do anything he can to stop the work of God. This was about to begin to happen.

I began to see persecution on unprecedented scales. At this point other unusual things happen. This outpouring of blessing and this outpouring of persecution began to be really stepped up and people were taken. I saw many penitentiaries all over the United States, especially concentrated in California. In the dream I saw many state prisons. In 1983 the Holy Spirit told me he was allowing the devil to build prisons in the state of California that would eventually become detention centers for Christians. These prisons were being built in rural areas that were normally 15 to 25 miles off of any main highway. I asked him why would that be the case and he said, "So that those people can be taken during the night hours."

Something spoke to me on the inside and said get out of here as fast as you can. I began to think to myself, "Oh my goodness, it is the end of the world now."

I began to run to my house as fast as I could. While I was running I heard in my spirit, the following scripture passage.

"It also forced all people, great and small, rich and poor, free and slave, to receive a mark on their right hands or on their foreheads, so that they could not buy or sell unless they had the mark, which is the name of the beast or the number of its name." (Revelation 13:16-17).

Now I am running back to my house as fast as I can because I am realizing my wife is there and she is alone. I reach for the doorknob and began to pull the door open. Another scripture came into my spirit at that time from Matthew 24, which says do not go back into your house.

So when you see standing in the holy place 'the abomination that causes desolation,' spoken of through the prophet Daniel – let the reader understand, then let those who are in Judea flee to the mountains. Let no one on the housetop go down to take anything out of the house. Let no one in the field go back to get their cloak. (Matthew 24:15-18)

I opened the door to see the most demonic presence I have ever seen. This presence that I encountered at the front door as I opened the door was very dark with a shroud of black around him. It was not his skin that was dark. It was a shroud of darkness around and over this being. This being was very sinister looking. Just his presence gripped my heart with great fear. At this point I began to scream as loud as I could.

Then I woke up from the dream. At that point the dream had been going for several hours. Later, I fell asleep again and instantly the dream started exactly where it left off.

I was facing this very sinister creature. It was very intense and it gripped my heart. I slammed the door and ran off. I realized that my wife was not in my home and that she was gone. I knew this by this presence.

It is hard to tell you how afraid I was. I have since learned how to combat demons so I am not afraid of that anymore. They don't usually go with me anymore because I know how to pray now. But at this time I had no comprehension about how to deal with this evil being.

So I began to run and I ran and I ran. In the dream I ran a couple of miles. I got caught by one of these strange looking police trucks. They knew my name even though I didn't tell them my name.

DETENTION CENTER

They took me to this government building. It was a large building. They took me into a room and there was my wife and this older gentleman that I began to call your evangelist. They had already been captured. These people knew exactly where to take me. That was very astounding to me because I wondered how do they know all of these things about people.

They began to politely interrogate us. They began asking us to be cooperative and to come into agreement with this new government and everything will be fine for you. Well, my wife is one of the boldest Christians I have ever met. She is also one of the kindest and most gentle believers I have ever met. But she will get in the devil's face. She and this older man began to preach to these people that were trying to convince us of this new alignment of this government.

So they took us out of that room and let us in another room. Now there was a lot of mind control interrogation. I could feel my mind being pulled into this new order. I began thinking if we just don't cause any trouble it will be okay. That is how my mind began to function. But the older gentleman and my wife began to fight this with all of their spiritual strength and they challenged it with scriptures. It was amazing to me because our capture was almost as though they had planned it out.

When we were being interrogated the mind control was phenomenal. It was not like any human being could do in an interrogation. My mind began to really be filled with anxiety and fear. Because my wife and this older gentleman kept being very bold and in your face with them they took us out of that room and into this very long corridor. In this corridor there were thousands of people lined up. The corridor seemed to be at least 100 yards long. It was probably longer than that. Every five or six minutes the people in this long line would take a step forward.

We had been in this line for a long time when people would barge in through the doors on the sides of the corridor and began to grill people and tell them to renounce their faith. They would never use the name Jesus. They would never use the name Jesus Christ. They would never use the name God. They would say you should renounce your faith in him while you can still live. They would say your faith is empty. It was a blasphemous challenge that these people were bringing against the people in the line. Every so often, someone in the line would crack. They would just collapse and these people would drag them away. They would renounce their faith in Christ.

It made me very uneasy to be in this line because I was not sure what they were going to do to us. I wondered if they were going to put us in prison or maybe beat us up.

Eventually we made it through a battery of three double doors. After going through the last double doors we were put into a holding cell. There was the old man in the front of the line and then my wife and then myself. They took this older gentleman into the room and closed the doors very quickly. I do not know what happened to him.

EXECUTION

About six minutes or so later they opened the doors wide open and what I saw made me experience the emptiest feeling I have ever experienced in my whole life. I saw this man that was very big. He was tall like a professional basketball player but was very big like a professional football player. He had a big satin hood over his head with eyeholes to see out.

My wife was in front of me and they began telling her she should renounce her faith and live. Now I realized what was happening because this man was standing there with a huge sword. It was a very frightening looking sword. I saw this table that was a little longer than the average human being and a little bit wider. My wife said she was not going to renounce her faith in Jesus. She began to preach to them powerfully. She began to rebuke the devil. They got angry and strapped her down on this table with her face up. This man was standing behind her with this sword. So he took the sword and chopped her head right off, right in my presence, I saw it.

This sword left an indelible mark in my life. Later I saw the same sword on the red cap worn by the Shriners. Their caps are called the Fez. They are red because they have promised to dip them in the blood of Christians. The sword is called the sword of scimitar.

I was more afraid of what was going to happen to me next than the fact that my wife had just died. I was more concerned about my life then with her dying. I was very afraid. I knew that I was going to die now. In my mind I knew that I was not going to make it. I was paralyzed. My mind began to torment me so much I almost literally blanked out. My stomach began to shout almost out loud asking Jesus to help me. The message could not get out because my mind was paralyzed. It was like I had the flu. My teeth were chattering and I was shaking with chills. I could not process my thoughts whatsoever. It was as though I had totally lost all faculties of my mind, my ability to cognitively be aware of what was going on. It was terrible. Although it only lasted for five or six minutes, it seemed like hours because of the extreme weight of this attack on me.

I began to really try to cry out to God from my stomach. Today I know it was my spirit crying out but in the dream it seems like a war in my stomach. Finally it was like something penetrated out of my stomach into my mind and I was able to spiritually call on Jesus and say, "I am afraid Jesus. Please save me. Help me."

At the very instant that communication happened, I felt a hand grip my shoulder. For a brief period of time I was actually more interested in the hand gripping me then I was in what was happening to me. As soon as this hand gripped me I got very warm and the chills left me. It was as though my mind could now see and I could comprehend clearly what was going on.

I will never forget the hand. It was a very rugged looking hand. It looked as though it had been through a great deal of work. It was almost like a man that is a blue-collar worker who uses his hands like a mechanic for a builder or a plumber. It was a very thick hand. It was a very solid hand. After a few moments I turned back and there was the Lord Jesus Christ standing behind me. All of a sudden he looked me in the eyes. He looked at me very sternly. It was not like a reproof or a conviction. It was more like he was just looking and peering into my life.

At the very moment I looked at him, his eyes were not brown or green or blue or anything like that. They appeared to be red like fire. They were looking clearly through my whole life. Somehow at that moment I was able to realize that him looking at me was actually looking through me. He knew everything about me. He knew my strengths. He knew my weaknesses. He knew every lie deep down inside of me. He knew every deception. He knew every place that I was afraid and that I had compartmentalized. By Him looking into me, my whole being was exposed to me. It was very frightening. It was a very intense moment. I wish I could say that seeing Jesus at that moment made me very happy. It did not. It made me very fearful. I understand now what the fear of the Lord is because of that experience.

A few moments after realizing my own depravity, He spoke to me. He looked sternly into my eyes and said, "Fear not my son for death will never hold you."

Instantly it was like courage flooded through me. I wish I could tell you that I got very bold and preached a great sermon and got everybody saved. But I didn't. It was just courage to go through what was before me.

I knew that he had saved me because of the prayer I prayed with the older man. When I looked at Him I knew He was the Lord of all and the king of every king. When I saw him I knew there was not one man that would not bow to him. There was not one tongue that would not confess him as Lord regardless of what side of the coin they are on when he reveals himself to them as Lord. Every knee will bow. This presence that he stood in was so powerful and so awesome and so anointed and so terrible that you knew there was no power on earth that could challenge Him.

Then these men strapped me down and they said, "You can renounce Him." I said, "No, I cannot renounce Him for He is the Lord of all and he should be your Lord."

That was my great sermon. I wish it could have been a lot longer but probably if I had got going with it I would have messed it up like I had messed up everything else.

When this man cut my head off I saw that as soon as it touched my neck, the moment the blade touched my neck, I was gone. I felt no death whatsoever. I was standing their holding a person's hand and I was looking down upon the scene. It was very grotesque. My head was cut off and I was bleeding profusely. Even though this hand was holding me up in the air I was actually more interested in seeing me dead then I was interested in the fact I had been delivered from death.

All of a sudden, I looked down and realized it was another one of these rugged hands holding my hand. I looked up and it was the Lord again. It was the Lord Jesus Christ.

At this point now it went from a stern, powerful, all-knowing God to a God who was holding my hand that gave me the understanding that now I was his son. I was his brother. He was not a shame to call me his brethren. All of a sudden, I had an understanding that I was equal with him. Not as God the deity or is Jesus the son of God, but as a son of God. I was not THE son of God as in the only begotten son of God, but I was a son of God. There was equality in the sense that we were brothers now. It was no longer a fearful thing for me to stand in his presence. There was immense acceptance. There was immense understanding. I had a clear understanding of things that I can preach now with great fire.

The Scripture says, "Precious in the sight of the Lord is the death of his saints." I know that when His children are coming to Him, it is precious to God.

When his children are coming through death it is precious to him. When he gathered me to himself in death, He showed me the scripture says it does not appear what we shall be but we know that when he appears we shall be like him. At that moment, I was like Jesus. I was like him in image, in faculty, in understanding. I could no longer see any of my weaknesses. None of my frailties were known to me any longer. I was completely delivered out of all of that. Truly to be in the presence of the Lord is to be like him.

All of a sudden, the man with the hood pulled off his hat and threw it down and said, "I will not kill another one of these people." With that the dream was over.

58

MY COMMENTS

Not only was Ken Peters not a believer at the time he received this dream, but he also did not know what the Bible says about the coming events of the last days. His ignorance of the scriptures makes him the ideal candidate to receive this dream because he had no agenda and no bias for or against any church doctrine. The church has been divided over questions about the timing of end-time events like the rapture of the church. But Ken Peters was wide open and simply reported what he saw in the dream exactly as it happened.

All prophetic words must be judged by the word. By that standard, I am not aware of any scriptures contradicted by this dream. Personally, I believe the dream is from the Lord and that it provides us insights that can help us gain a better understanding of the scriptures than we would otherwise have. I gained the following insights from this dream:

1. THE RAPTURE:

Christians are divided regarding the timing of the rapture. Some believe it happens before the tribulation (pre-trib). Others believe it happens in the middle of the tribulation (mid-trib). Still others believe it happens at the end of the tribulation period (post-trib). Ken Peters only saw the resurrection of believers who were already in the grave. He did not see the rapture of any living person. However, just because he did not see that event does not mean the living believers were not also taken at the same time. A careful reading of 1 Thessalonians 4:18 reveals that the living believers will be caught up together with the dead in Christ in the clouds to meet the Lord in the air.

But we do not want you to be uninformed, brethren, about those who are asleep, so that you will not grieve as do the rest who have no hope. For if we believe that Jesus died and rose again, even so God will bring with Him those who have fallen asleep in Jesus. For this we say to you by the word of the Lord, that we who are alive and remain until the coming of the Lord, will not precede those who have fallen asleep. For the Lord Himself will descend from heaven with a shout, with the voice of the archangel and with the trumpet of God, and the dead in Christ will rise first. Then we who are alive and remain **will be caught up together with them in the clouds** to meet the Lord in the air, and so we shall always be with the Lord. Therefore comfort one another with these words. (1 Thessalonians 4:13-18 NASB)

Since the living are "caught up together with them" we can conclude both groups must be taken during the same rapture. Otherwise, they would not be able to meet together in the clouds. The Lord brought with Him those believers who had already died because their spirits had gone to heaven without their bodies. They have been living in heavenly bodies while their earthly bodies are still in the grave. At the rapture, their bodies will come up out of their graves all over the world. Their spirits will then move back into their earthly bodies, which will be re-created by God at that time. So then both groups, the living and the dead, will finally be together, both living in heaven in their earthly bodies, which will be changed by God at the rapture to make them immortal and imperishable (1 Corinthians 15:51-57). Their bodies will no longer be decaying day by day (2 Corinthians 4:16). It is also interesting that Ken Peters described them in their new bodies as being brighter than the sun.

2. THOSE WHO ARE LEFT BEHIND:
Among those who are left behind, many will realize what happened and they will repent and get saved. That is exactly what Ken Peters saw when he encountered the older gentleman. Those believers will be rounded up and taken to detention centers where they will be told they must accept the mark of the beast or be killed by the antichrist. Many will refuse the mark and will be martyred, just as he saw in the dream. It will be very unpleasant for those who are left behind, but they can still be saved by turning to the Lord. In fact, in the dream Ken Peters himself was one of those people who was saved after the rapture and was beheaded for his faith. We see in Revelation 20 these people are part of the first resurrection and they will reign with Christ for a thousand years on the earth.

I saw thrones on which were seated those who had been given authority to judge. And I saw the souls of those who had been beheaded because of their testimony about Jesus and because of the word of God. They had not worshiped the beast or its image and had not received its mark on their foreheads or their hands. They came to life and reigned with Christ a thousand years. (The rest of the dead did not come to life until the thousand years were ended.) This is the first resurrection. (Revelation 20:4-5). Those who are left behind will endure great trials, but God gives special blessings to those who endure to the end.

Blessed and holy is the one who has a part in the first resurrection; over these the second death has no power, but they will be priests of God and of Christ and will reign with Him for a thousand years. (Revelation 20:6)

They will live on the earth during the thousand year reign of Christ. There will continue to be mortal human beings living on earth with them during that time, but those who overcame the onslaught of the antichrist will be ruling and reigning with Christ and living in immortal, imperishable bodies.

3. THE TRUMPET BLAST:
The car horn that Ken Peters heard at the start of the dream was the trumpet blast that coincides with the resurrection of the dead in 1 Thessalonians 4:13-18.

4. FAMINES AND EARTHQUAKES:
The famines and earthquakes described in the dream are scriptural. Matthew 24:7 specifically tells us there will be famines and earthquakes in various places.

5. THE EXECUTIONS:
The gruesome execution method described in the dream is very much in line with scripture. Revelation 4:20 specifically tells us there will be many

believers beheaded during the tribulation because of their testimony of Jesus and because of the word of God.

I did not find anything contrary to scripture in any part of this dream. In summary, it helped me to see the sequence of events and the nature of the events more clearly. I thank God for this revelation and I hope it has helped you also!

http://Z3news.com

FIVE PROOFS

April 24, 2015 9:58 pm By James Bailey

Here are five reasons why I believe the rapture comes before the wrath of God is poured out at the end of this age.

1. GOD HAS PROMISED TO DELIVER US

The wrath of God is not for innocent people, but for unrepentant sinners who have rejected him. Everyone who has received Jesus Christ and turned away from sin is innocent before God because the blood of Jesus the Messiah paid their debts in full. So God has promised to deliver His people from the wrath to come:

Much more then, having now been justified by His blood, we shall be saved from the wrath of God through Him. (Romans 5:9)

Wait for His Son from heaven, whom he raised from the dead, that is Jesus, who delivers us from the wrath to come. (1 Thess 1:10)

For God has not destined us to wrath, but for obtaining salvation through our Lord Yeshua the Messiah, who died for us, so that whether we are awake or asleep, we live together with Him. (1 Thess 5:9-10)

Other scriptures reveal the way he will fulfill the above promises is through the rapture, gathering his people together to meet the Lord in the air.

For the Lord himself, with a cry of command, with the archangel's call and with the sound of God's trumpet, will descend from heaven, and the dead in Christ will rise first. Then we who are alive, who are left, will be caught up in the clouds together with them to meet the Lord in the air; and so we will be with the Lord forever. Therefore encourage one another with these words. (1 Thessalonians 4:16-18)

2. GOD PROTECTS THE 144,000 BOND-SERVANTS

During the outpouring of God's wrath, the fifth trumpet releases locusts who are instructed to harm everyone on earth except the 144,000 who have been sealed by God. Since we know the seal is only given to the 144,000 then we also know the rest of God's people must already be removed from the earth through the rapture. Otherwise, God would be instructing the locusts to harm his own people, which would be a violation of his promise to deliver them from his wrath. Instead we see God and his angels consistently taking actions to protect all of his people. This provides further evidence that the rapture happens before the wrath of God is poured out.

While most of his people are completely removed from the earth in the rapture, this group of 144,000 Jewish believers, called bond-servants, remains on earth during the great tribulation. 12,000 come from each of the 12 tribes of Israel. Consistent with the promises shown above, this group will also be delivered from God's wrath by a supernatural seal placed on their foreheads.

After the rest of God's people are delivered in the rapture, four angels are given power to harm the earth, but they are told not to release any harm until these 144,000 bond-servants are protected by the seal. Another angel with the seal of God says to them, "Do not harm the earth or the sea or the trees until we have sealed the bond-servants of our God on their foreheads" (Revelation 7:1-3).

The following instructions are given to the locusts:

They were told not to hurt the grass of the earth, nor any green thing, nor any tree, but only the men who do not have the seal of God on their foreheads. And they were not permitted to kill anyone, but to torment for five months; and their torment was like the torment of a scorpion when it stings a man. And in those days men will seek death and will not find it; they will long to die, and death flees from them. (Revelation 9:4-6)

3. GOD WILL CUT SHORT THOSE DAYS

Jesus promised those days would be cut short for the sake of His people, the chosen ones, also called the elect.

Unless those days had been cut short, no life would have been saved; but for the sake of the chosen ones those days will be cut short. (Matthew 24:22)

The Lord allows the antichrist to reign over the earth for 42 months. After that his reign is cut short, interrupted by the day of the Lord when the Lord returns to rescue his people from the wrath to come. This promise to cut short those days is further evidence the rapture happens before the wrath of God.

4. THE RAPTURED CHURCH IS BEFORE THE THRONE PRIOR TO THE WRATH

The sequence of events in Revelation 6-8 reveals God's people are rescued by the rapture just before the wrath of God is poured out. The appearance of the raptured believers standing before the throne in Revelation 7 proves they are rescued before the wrath of God begins in Revelation 8. The first six seals are broken in Revelation 6 releasing trouble in the earth. Then John saw the raptured church standing before the throne in heaven in Revelation 7. Then in Revelation 8 the seventh seal is broken and the seven angels are given seven trumpets, which marks the beginning of the outpouring of the wrath of God. The sequence of these scriptures is not a coincidence or a mistake. This is further evidence of God's plan to deliver His people from the coming wrath of God through the rapture.

In Revelation chapter 7, John saw the raptured church standing before the throne of God.

After these things I looked, and behold, a great multitude, which no one could count, from every nation and all tribes and peoples and tongues, standing before the throne and before the Lamb, clothed in white robes, and palm branches were in their hands; and they cry out with a loud voice, saying, "Salvation to our God who sits on the throne, and to the Lamb."

And all the angels were standing around the throne and around the elders and the four living creatures; and they fell on their faces before the throne and worshiped God, saying, "Amen, blessing and glory and wisdom and thanksgiving and honor and power and might, be to our God forever and ever. Amen."

Then one of the elders answered, saying to me, "These who are clothed in the white robes, who are they, and where have they come from?" I said to him, "My lord, you know."

And he said to me, "These are the ones who come out of the great tribulation, and they have washed their robes and made them white in the blood of the Lamb. For this reason, they are before the throne of God; and they serve Him day and night in His temple; and He who sits on the throne will spread His tabernacle over them. They will hunger no longer, nor thirst anymore; nor will the sun beat down on them, nor any heat; for the Lamb in the center of the throne will be their shepherd, and will guide them to springs of the water of life; and God will wipe every tear from their eyes." (Revelation 7:9-17)

We are not told specifically how this great multitude exited the earth to get to heaven, but we are told they were here during the great tribulation because the elder told John they came out of the great tribulation. We are even told the things they suffered, including hunger, thirst, exposure to heat, and the shedding of tears. Given the large number of them and the fact they all come out of the earth during the same short period of time, it is not likely they all died of natural causes. It is also not likely they were martyred because when martyrs are killed in other passages they are honored by recognition of their great sacrifice. If this great multitude also laid down their lives failing to recognize that would be disrespectful and inconsistent with other scriptures. Therefore, we can conclude this great multitude left the earth in the rapture.

5. RAPTURE HAPPENS ON THE DAY OF THE LORD

This point was so long it required a separate post. Don't miss it because it is loaded with important points! See here: Eight Proofs the Rapture Happens

On the Day of the Lord (http://z3news.com/w/proof-the-rapture-happens-on-the-day-of-the-lord).

CONCLUSION

In conclusion, there are five reasons why we know God will protect His people from the wrath to come:

1. He promised us he would deliver his people from wrath.

2. He releases locusts to harm everyone except for the 144,000 bond-servants, proving the rest of his people are already removed from the earth.

3. He promised to cut short the tribulation for the sake of His chosen ones.

4. Prior to the wrath being poured out in Revelation 8, John saw the raptured church standing before the throne of God in Revelation 7.

5. The rapture happens on the day of the Lord just before the wrath of God is poured out.

AUTHOR: JAMES BAILEY

James Bailey is an author, business owner, husband and father of two children. His vision is to broadcast the good news of Jesus Christ through blog sites and other media outlets.

THE FULFILLMENT OF REVELATION 12 –
A FINAL PROOF

A once-in-a-century heavenly sign appears on September 23 2017. It is a harbinger of the mid-point of the Tribulation.

REVELATION 12:1-2
Now a great sign appeared in heaven: a woman clothed with the sun, with the moon under her feet, and on her head a garland of twelve stars. Then being with child, she cried out in labor and in pain to give birth.

Revelation 12 is the sign of the woman in Israel, and the twelve stars are the tribes of Israel.

This is a sign of major import. The dragon (Satan) is thrown from heaven and about to persecute Israel for the remainder (3.5 years of the tribulation) of time left in the 70th week of Daniel. The sign is divided into several parts. First, the woman clothed with the sun. She is represented by Virgo (the virgin). The moon is under the feet of Virgo. The last part of September of 2017 places the moon correctly. Above the constellation Virgo is Leo. It has nine primary stars, and three wandering stars, or a total of twelve – again a perfect combination. A garland of twelve stars is thus created.

This is so rare if you search 150 years before the event, and 150 years later, you can find no other matching results. The final piece of the cryptographic puzzle is Jupiter. It enters Virgo in August of 2016, and spends 400 days there. **On September 23, 2017, Jupiter is right in the "womb" of Virgo. A birth is imminent. The world is about to witness the birth of the Messianic Kingdom.** Only God Himself could have set up this sign in the Mazzeroth or heavens, and pre-ordained it in the Book of Revelation. **It is a marker to the mid-tribulation.** This sign is so rare it is thus unmistakable that 2017 represents one of the Tribulation years, right before the Great Tribulation or the last 3.5 years.

PLANET X AND THE MYSTERIOUS DEATH OF DR. ROBERT HARRINGTON

YOWUSA.COM, 22-May-2008

John DiNardo

Janice Manning

Dr. Robert S. Harrington, the chief astronomer of the U.S. Naval Observatory, died before he could publicize the fact that Planet X is approaching our Solar System.

Many feel his death part of a cover-up. One in which government agencies quickly moved to conceal the most earth-shaking discovery in history. If so, the search for truth begins in New Zealand.

In 1991, Dr. Robert S. Harrington, the chief astronomer of the U.S. Naval Observatory, took an 8-inch telescope to Black Birch, New Zealand, one of the few viewing points on Earth optimal for sighting Planet X, which he definitively calculated to be approaching from below the ecliptic at an angle of 40 degrees.

However, the source below quotes Dr. Harrington as predicting 30 degrees, not 40.

The Independent, September 18, 1990

Dr Harrington says the most remarkable feature predicted for Planet X is that its orbit is tilted 30 degrees away from the ecliptic, the main plane of the solar system, where all previous searches have concentrated. His models also predict a greater distance from the Sun, about 10 billion miles, or between two or three times as distant as Pluto.

By analyzing time-lapse photographs using the "blink comparison" technique, originated by famed Pluto discoverer, Clyde Tombaugh, Dr. Harrington proved that Planet X was indeed inbound into our Solar System. Harrington sent back reports of this ominous discovery, but died of what was reported to be esophageal cancer before he could pack up his telescope and come home to hold what would have been a highly publicized press conference.

The source below indicates that Harrington began the search at Black Birch in April, 1991, two years before his death. However, considering the painstaking process of trying to find a distant needle in a very large haystack, two years is not much time at all.

The Independent, September 18, 1990

In April (1991) the new sweep starts in earnest at the Black Birch Observatory in New Zealand. A modest 8-inch telescope, similar to that used by Mr. Tombaugh, will examine the northern part of the constellation Centaurus. Pairs of photographs of the same region of sky taken on successive nights will be sent to Washington. Using a blink comparator, a device that compares two photographs, Dr Harrington hopes to locate any faint object that has moved during the interval between the two pictures.

Weigh together the following numerous facts:

1. Publicity of Harrington's discovery would pose a great threat to the social, political, and economic stability of the World;

2. It is extremely unlikely that a man would contract cancer and die within a matter of days, suffering the sudden impact of pain and debilitation characteristic of rapid onset cancer, while miraculously traveling on a physically demanding expedition and performing intensive astronomical operations. From his obituary:

Robert (Bob) Harrington died on Jan. 23, 1993 after a short, but determined, battle against esophageal cancer. He left his wife, Betty, two daughters, a sister, and his parents.

3. Even though Dr. Harrington died before he could publicize the fact that Planet X is approaching our Solar System, the U.S. Naval Observatory was apprised of it, as was NASA, yet these government agencies concealed this most earth-shaking discovery in history; (This appears to be true, as this writer spent a week searching the Internet for information to the contrary and found none.)

4. Furthermore, in publishing Dr. Harrington's obituary, the U.S. Naval Observatory went out of its way to gratuitously lie about Dr. Harrington's final achievement, stating that "in his final years, Dr. Harrington had lost interest in the" [two-century astronomical] "search for Planet X." The obituary had the following to say about Planet X:

    Considerations on the stability of the solar system led Bob to collaborate with T.C. Van Flandern in studies of the dynamical evolution of its satellites, and to an eventual search for "Planet X", conjectured to lie beyond Pluto and to be responsible for small, unexplained, residuals in the orbits of Uranus and Neptune. Late in his career Bob seemed quite skeptical of such an object, however.

5. Secrecy about the approach of Planet X could have been much better achieved by NOT mentioning Harrington's interests in his later years; so why did they concoct this lie gratuitously? Because they are not very clever, this brain trust ensconced within the bowels of the national intelligence apparatus.

6. Dr. Harrington's colleague in the search for Planet X, Dr. Tom Van Flandern reversed his affirmative statements about the approach of Planet X and became peculiarly silent on the issue. In Meta Research Bulletin (September 1999), he states:

"Three more trans-Neptunian objects confirm the presence of a second asteroid belt in the region beyond Neptune. This probably indicates that the hypothetical Planet X is now an asteroid belt rather than an intact planet."

7. As long ago as Dec. 30, 1983, the Washington Post published a front-page article announcing the discovery, by the Infrared Astronomical Satellite, of this very same celestial object, calling it "a heavenly body, possibly as large as the giant planet Jupiter, and possibly so close to Earth that it would be part of our solar system"; yet, this great scientific heralding was immediately reversed as the news media fell astoundingly silent, and has been silent for the past quarter-century. Here is a clip from that report.

Washington Post, December 30, 1983

The mystery body was seen twice by the infrared satellite as it scanned the northern sky from last January to November, when the satellite ran out of the supercold helium that allowed its telescope to see the coldest bodies in the heavens. The second observation took place six months after the first and suggested the mystery body had not moved from its spot in the sky near the western edge of the constellation Orion in that time. "This suggests it is not a comet because a comet would not be as large as the one we've observed and a comet would probably have moved," Houck said. "A planet may have moved if it were as close as 50 billion miles but it could still be a more distant planet and not have moved in six months time."

If a massive planet and its entourage is approaching our Solar System one quarter-century ago, and the media chose to conceal it, does this wrecking ball suddenly become a soap bubble? Or does the fear-bred silence of the powers-that-be suggest that they are desperately protecting their empire of social, political, and economic domination for as long as possible before they duck into their elaborate underground cities in a wishful attempt to survive God's promised wrath, now hurtling toward us all?

NASA AND THE IRAS SATELLITE –
THE DISCOVERY OF PLANET X

Why did NASA shut down IRAS, the infrared telescope?

From an Insider: Planet X had been imaged by NASA's IRAS infrared sensing satellite in 1983 and the mechanical failure story was used as a cover story. Once the IRAS data started pouring in, that's when they found Planet X. It was approaching from south of the ecliptic.

This was not good news because the majority of the world's observatories are north of the Equator, and the decision was made to devote the remaining lifespan of the IRAS spacecraft to the observation of this one object.

After releasing the mechanical failure cover story, controllers used IRAS's remaining fuel to maintain a constant track on the object until control over the spacecraft was entirely lost.

THE SOUTH POLE TELESCOPE (SPT)

The decision was then made to spend billions of dollars to construct a telescope at the South Pole. It would be far more powerful and survivable than the 1983 IRAS spacecraft and Hubble Space Telescope put together, and this manned observatory would begin tracking Planet X / Nibiru from the skies of Antarctica.

Why else would a telescope requiring huge expenditures of logistics and funds be built in such a desolate location? Why not Chile, for example!

It was reported in 2001 that $2.2 Trillion was missing from the Pentagon's budget. There are reports (one of them from Jesse Ventura's show) about a Deep Underground Military Base below Denver International Airport. It is highly likely this "missing money" was used to construct this, and a multitude of other military bases in anticipation of the arrival of Nibiru.

I hope at this point it's obvious – they've known about this object for years.

In 2011 deaths of electromagnetically sensitive animals, birds and fish started to occur and escalate worldwide. This is due to the release of methane gas from the earth's core. People began to notice sudden and erratic fluctuations in their compasses. Then on 11 January 2011, the increased infusion of Nibiru's reverse polarity magnetism caused Earth's rotational axis to shift slightly, making the Sun rise two days early in Greenland.

NIBIRU

Nibiru (aka Planet X and referred to by the Vatican as Wormwood) has a diameter of 179,028 to 183,191 kilometers. Its mass is 3.34 times the mass of Jupiter. The object is huge, but it is a red dwarf star visible only in the infrared spectrum. Its composition is cesium, iron oxide, iron, oxygen, and ozone. It has seven planets (moons) orbiting it and innumerable asteroids. Its forward velocity is 25,890 mph. Its rotation is clockwise. It has an outer magnetic field of 16.326 AU in diameter. Its inner magnetic field is .184 AU in diameter. Its core magnetic field is .066 AU in diameter. The Vatican is so interested in this object they've funded and participate in an observatory in a remote part of Arizona, where they watch the object approach the earth. The Wormwood terminology is from the Book of Revelation (the Apocalypse).

Planet X is not going to destroy the earth but it's going to affect for a period of years every living person on the planet. It's going to cause massive destruction. Governments don't spend several trillion dollars without a very good reason.

Planet X is the reason we have the Seed Vault in Greenland.

So much of what we've learned is complex and some of it may be new to you. Let's go over what we know so far. And very importantly – do your own research! Let this be a guide for you. Virtually everything we've gone over so far can be validated through research. You can research on the Internet. You can buy an 8-inch Meade scope, set it up (though usually setup is a two-month project to become proficient) and you can see it for yourself.

You have to think like an intelligence officer to understand how the government operates. They operate in terms of covers and legends when they don't want to disclose their real purpose. In terms of the South Pole Telescope, for instance, a good cover is "We need to study cosmic rays in an undiluted atmosphere in order to understand the age of the earth." Sounds ridiculous, doesn't it? Well, that's one of their supposed reasons for the SPT.

You have to use your intuition as well as rational thought. You have to train yourself in critical thinking and analysis. Buy and read a good book on the Intel community and how they operate. Remember, the government(s) have taken care of their own. You have to take care of you. Russia is just finishing 5,000 shelters in Moscow. They're much more open (in this regard anyway) on this topic than other governments.

I personally estimate the US probably has over 150 Deep Underground Military Bases (in every state). Russia has some that can accommodate over 100,000 people (at the base of mountains). These are entire underground cities that have been under construction since the 1980s.

Until the discovery of Planet X astronomers had regarded the writings of the ancient Sumerians about this object as legend. When Planet X was discovered in 1983 they suddenly learned that the Sumerians were not the primitive people they have been made out to be by "intellectuals" of today.

What these modern scientists discovered is that Planet X is a huge DARK planet which is much, much larger than the earth. It is so large we refer to it as a planet, yet it has an elliptical orbit more like that of a comet than a planet. Its orbit is so elongated that it spends most of its time in the darkness of outer space and it comes back into our inner solar system every 3,630 years!

Because it is so large, Planet X also has an electromagnetic field surrounding it that is second only to that of the Sun in the solar system. In other words, it acts as a powerful magnet in space, pushing all kinds of space debris out in front of it as it advances through space, and trailing a million miles or more long tail of space debris behind it.

The astronomers and scientists who lived in 1983 also learned that Planet X was INBOUND to our solar system. In view of the gigantic size of Planet X and its powerful electromagnetic field (which will cause it to wrest magnetic polarity of the earth from the sun), the scientists knew that when it approaches the earth it will cause all kinds of destruction to each of the planets of the solar system.

PIONEER

When Pioneer 10 was sent out to study Planet X it encountered a mysterious force – dubbed the Pioneer anomaly by baffled NASA scientists. It was gently pushed against (by the force of the magnetosphere of Planet X) and moved in the opposite direction back towards the Sun. Its forward momentum was decelerated.

Pioneer was travelling towards the right of Nibiru at this point. This tells us two things: (1) that Nibiru's outer magnetic force is so strong it can affect objects 46.8 AU from Nibiru (Pioneer's approximate distance at the time), way beyond the central and certainly much stronger magnetic field that we've calculated spans 16AU with a reach or radius of 8AU; (2) that Nibiru's magnetism rotates in a clockwise fashion, because that is the only way it could push Pioneer back when the deep space probe was flying to the right of it. If it rotated anticlockwise and the Pioneer 10 was travelling in the same direction it would push the probe slightly forward and to the right of its trajectory. The effect would have been the opposite as the probe would have been seen to accelerate, not decelerate, in its drift.

I'm grateful for these conclusions to a scientist who must remain anonymous who has related the data (which was at one time privately published).

Let's get back to summarizing in simple terms – the far greater electromagnetic fields of Planet X will cause great earthquakes (greater than the highest level of 9.5 known to date), volcanic eruptions, tsunamis and a shift of the magnetic poles.

Now it is public knowledge that government agencies are on a need-to-know classification (intra-agency), and that they never publish information in advance that could cause an economic or other disruption. Enough said.

However, if you do any research at all you will run into different levels of disinformation. Some of this is from uneducated members of the public, and some of it is very likely on purpose. You have to use your head when you do research.

These underground sanctuaries that have been built are stocked with food and water, gold and silver bars, wrapped currency, medicines and whatever it takes to live in comfort when they are entered. Why? If Planet X did not exist there would be no need to build them.

HUBBLE

The real reason for the Hubble Space Telescope was NOT to explore the solar system and the Universe, as some claim from time to time. The real reason for spending billions of dollars to construct the telescope, place it in orbit, and then immediately replace parts on it was for the purpose of viewing inbound Planet X.

THE VATICAN

Even the Vatican hierarchy in Rome have their hands in this. They have spent millions of dollars quietly building one of the most technologically advanced observatories in southern Arizona specifically for the purpose of viewing and tracking Planet X on its inbound journey to our solar system. They have deliberately kept their involvement in this enterprise quiet. Some sources have gone to the trouble of searching out the location of this observatory in the mountains of Arizona, and even photographing it, so its existence is carefully documented. But why the secrecy? The Vatican hierarchy in Rome regards Planet X as the prophesied and much dreaded "Wormwood" of Revelation 8:10-11. They know it is coming, and they are spending millions of dollars to track it on its inbound journey.

THE EMP EFFECT

What if an X-60 solar flare leaps from the sun when Planet X approaches? The coronal mass ejection (CME) heads straight toward Earth. What will happen? How much warning will we have?

This type of event may be our first overt sign of the nearness of the approach of Planet X. The CME is a slow-moving cloud of charged particles. It is accompanied by an X-ray burst. Both have devastating effects. The X-ray burst travels at the speed of light and would reach Earth's surface in eight minutes.

The X-rays would affect all Earth-orbiting satellites in line of sight. GPS communications would go down, and so would communications satellites. That would be your first warning and an obvious signal to those watching.

This would be a one-two punch. The second would be the CME (arriving 3-4 days later) that would cause transformers (which operate with copper wiring) to heat up and overload. As design capacity is overloaded, they burn up.

Most reactors require electricity to operate the cooling systems. Huge levels of radioactive fuel are on-site. Cooling on a continuous basis is the only way to prevent a meltdown.

This apocalyptic scenario is probable, if not guaranteed. If the reactor cores are not continuously cooled, a catastrophic reactor core meltdown and fires in storage ponds for spent fuel rods is the imminent result.

With a widespread grid collapse, in just hours after the backup generators fail or run short of fuel, the reactor cores melt down. Within a couple of days without electricity, the water bath over the spent fuel rods will "boil away" and the stored fuel rods will melt down and burn.

Transformers are made to order and custom-designed for each installation. They weigh as much as 300 tons and cost more than $1 million. Given that there is currently a three-year waiting list for a single transformer (due to recent demand from China and India, lead times have grown to three years), you can begin to grasp the implications of widespread transformer losses.

The Nuclear Regulatory Commission only requires one week's supply of backup generator fuel at each reactor site. The public will have one week to prepare for Armageddon.

ELECTROMAGNETIC PULSE ATTACK

According to the 2004 Commission to Assess the Threat to the United States of EMP Attack *(Executive Report)*, "Several potential adversaries have or can acquire the capability to attack the United States with a high-altitude nuclear weapon-generated electromagnetic pulse (EMP). A determined adversary can achieve an EMP attack capability without having a high level of sophistication."

It goes on to briefly address the effects, "EMP is one of a small number of threats that can hold our society at risk of catastrophic consequences. EMP will cover the wide geographic region within line of sight to the nuclear weapon. It has the capability to produce significant damage to critical infrastructures and thus to the very fabric of US society..." The Commission's chairman has testified that within one year of such an attack, 70%-90% of Americans would be dead from such causes as disease and violence. It is also highly plausible that many Americans would die of starvation due to the interruption of the national food supply.

According to the Washington Department of Health, Office of Radiation Protection, "A 1.4 Megaton bomb launched about 250 miles above Kansas would destroy most of the electronics that were not protected in the entire Continental United States."

So, as you can see, both a massive solar storm and an EMP event could quite possibly end civilization as we know it. I know that sounds drastic, but in the

United States and other technologically advanced countries, how would the mass population handle a prolonged event with very little or quite possibly, no electricity? As the Commission noted, our society is utterly dependent on our electrical grid for everything.

Trucking and transportation
Gas stations and refineries
Information and communications
Commercial production of food and goods
Water purification and delivery
Most of our military capability

These are only a handful of things that we take for granted because they are always there. If the gas stations were out of order, and no refineries able to produce more fuel, can you imagine how quickly our "civilized society" would break down? With that event alone, grocery store shelves become empty within a matter of days and farmers can't transport any goods. If you were not aware, grocery stores do not stock much extra produce or food "in the back of the store." In order to maintain a high profit margin, stores maintain only a few days worth of staples until another shipment arrives. This not only conserves space, but allows for them to keep their overhead lower, among other things.

Once the gas stops flowing and the shelves are wiped clean, how long will your neighbor remain *civil*?

Several tests and scenarios have shown that cell phones will be one of the first tell-tale signs of an electromagnetic event because of the enormous percentage of the population carrying one. If the power grid were to simply go down, this wouldn't affect your cellphone. Depending on your location, your local cell towers probably have back-up power systems, as well. However, the cell towers, backup power and your cell phone will all be disabled after an electromagnetic event, offering you a clue as to what has just happened.

The Commission went on to assess just how our society would be impacted from an EMP event, including how well cars and trucks can handle the burst of electromagnetic waves.

THE AUTOMOBILE AND TRUCKING INFRASTRUCTURES
{brief excerpt from the Commission's 2008 report}

"Over the past century, our society and economy have developed in tandem with the automobile and trucking industries. As a consequence, we have become highly dependent on these infrastructures for maintaining our way of life.

Our land-use patterns, in particular, have been enabled by the automobile and trucking infrastructures. Distances between suburban housing developments, shopping centers, schools, and employment centers enforce a high dependence on the automobile. Suburbanites need their cars to get food from the grocery store, go to work, shop, obtain medical care, and myriad other activities of daily life. Rural Americans are just as dependent on automobiles, if not more so. Their needs are similar to those of suburbanites, and travel distances are greater. To the extent that city dwellers rely on available mass transit, they are less dependent on personal automobiles. But mass transit has been largely supplanted by automobiles, except in a few of our largest cities.

As much as automobiles are important to maintaining our way of life, our very lives are dependent on the trucking industry. The heavy concentration of our population in urban and suburban areas has been enabled by the ability to continuously supply food from farms and processing centers far removed. As noted above, cities typically have a food supply of only several days available on grocery shelves for their customers.

Replenishment of that food supply depends on a continuous flow of trucks from food processing centers to food distribution centers to warehouses and to grocery stores and restaurants. If urban food supply flow is substantially interrupted for an extended period of time, hunger and mass evacuation, even starvation and anarchy, could result.

Trucks also deliver other essentials. Fuel delivered to metropolitan areas through pipelines is not accessible to the public until it is distributed by tanker trucks to gas stations.

Garbage removal, utility repair operations, fire equipment, and numerous other services are delivered using specially outfitted trucks. Nearly 80 percent of all manufactured goods at some point in the chain from manufacturer to consumer are transported by truck.

The consequences of an EMP attack OR A Y-CLASS SOLAR FLARE (being the most extreme and most likely) on the automobile and trucking infrastructures would differ for the first day or so and in the longer term. Either will certainly immediately disable a SIGNIFICANT portion of the 130 million cars and 90 million trucks in operation in the United States. Vehicles disabled while operating on the road can be expected to cause accidents. With modern traffic patterns, even a very small number of disabled vehicles or accidents can cause debilitating traffic jams. Moreover, failure of electronically based traffic control signals will exacerbate traffic congestion in metropolitan areas. Y-Class Solar Flares can be much more dangerous than strictly an EMP stand-alone event.

VULNERABILITY OF THE AUTOMOBILE AND TRUCKING INFRASTRUCTURES

The Commission tested the EMP susceptibility of traffic light controllers, automobiles and trucks.

The summary of the tests conclude that traffic light controllers will begin to malfunction following exposure to EMP fields as low as a few kV/m, thereby causing traffic congestion.

For automobiles, approximately 10% of the vehicles on the road will stop, at least temporarily, thereby possibly triggering accidents, as well as congestion, at field levels above 25k V/m. For vehicles that were turned off during the testing, none suffered serious effects and were able to be started.

Of the trucks that were not running during EMP exposure, none were subsequently affected during the test. Thirteen of the 18 trucks exhibited a response while running. Most seriously, three of the truck motors stopped. Two could be restarted immediately, but one required towing to a garage for repair. The other 10 trucks that responded exhibited relatively minor temporary responses that did not require driver intervention to correct. Five of the 18 trucks tested did not exhibit any anomalous response up to field strengths of approximately 50k V/m.

In regards to the airline industry, "Although commercial aircraft have proven EM protection against naturally occurring EM environments [such as lightning], we cannot confirm safety of flight following [severe or hostile] EMP exposure. Moreover, if the complex air traffic control system is damaged by EMP, restoration of full services could take months or longer."

It should be clearly noted and understood that these tests only involved pulses up to 50k V/m. Russia, the United States and several other countries are speculated to have weapons that can produce 100-200k V/m in purpose-built EMP warheads. One of the early tests in nuclear detonation and EMP study was called Starfish Prime. This event knocked out streetlights as far away as Hawaii and was rated at a measly 5.6k V/m.

Another potential problem, this time with solar storms, is that EMPs last only milliseconds and can be presumed to be rated no more than the typical 50k V/m. Coronal mass ejections from the Sun and X & Y-Flares can last for several minutes and have dire consequences should it hit the Earth. EMPs are also rather regionalized. One single EMP could affect the entire continental United States, but we might be able to rely on Britain or other countries to help us out. During a large enough solar storm, it might very well engulf the entire planet as the magnetic field surrounds the Earth.

CONCLUSIONS

In conclusion, depending on the technology used, you have a decent chance that should an EMP or solar storm occur while you are driving home from work, you will be able to make it home as long as you are careful to avoid collisions. It boils down to terrain, distance and strength. Once home, however, is an entirely different story!

There will be no more fuel available. There will be no more food and water for purchase. There will be no more iPhone or internet. And if you do find these things, what will be the price? Your dollars will very likely mean nothing to anyone with common sense. The art of bartering will very quickly take on a new importance for your own survival.

If this event were to occur, you could count on a very prolonged period of great civil unrest, riots, theft and wide spread violence. Repairs will be very slow and new parts for the large generators and power plants will likely have to be manufactured overseas and delivered to the United States. Furthermore, these foreign factories would have to retool their machines to create the specific part that we need if they are not already our supplier. And that is *if* the other industrialized nations aren't affected, as well.

http://www.truthistreason.net/emp-attack-and-solar-storms-the-complete-guide

Courtesy of Kevin Hayden

Founder, TruthisTreason.net

TSUNAMIS, FOOD STORAGE
& WATER PURIFICATION

A tsunami is a series of waves that may be dangerous and destructive. When you hear a tsunami warning, move at once to higher ground and stay there until local authorities say it is safe to return home.

BEFORE

Find out if your home is in a danger area.

Know the height of your street above sea level and the distance of your street from the coast. Evacuation orders may be based on these numbers.

Be familiar with the tsunami warning signs.

Because tsunamis can be caused by an underwater disturbance or an earthquake, people living along the coast should consider an earthquake or a sizable ground rumbling as a warning signal. A noticeable rapid rise or fall in coastal waters is also a sign that a tsunami is approaching.

Make evacuation plans.

Pick an inland location that is elevated. After an earthquake or other natural disaster, roads in and out of the vicinity may be blocked, so pick more than one evacuation route.

Have disaster supplies on hand.

- Flashlight and extra batteries

- Portable, battery-operated radio and extra batteries

- First aid kit and manual

- Emergency food and water

- Nonelectric can opener

- Essential medicines

- Cash and credit cards

Check food supplies and test drinking water.

Fresh food that has come in contact with flood waters may be contaminated and should be thrown out. Have tap water tested by the local health department.

HEALTH EFFECTS OF TSUNAMIS

Immediate health concerns:

- After the rescue of survivors, the primary public health concerns are clean drinking water, food, shelter, and medical care for injuries.

- Flood waters can pose health risks such as contaminated water and food supplies.

- Loss of shelter leaves people vulnerable to insect exposure, heat, and other environmental hazards.

- Medical care is critical in areas where little medical care exists.

When you are picking out the items to stock your emergency larder with, you want to make certain that you have a good assortment of different foods so that your diet is balanced and you can avoid the mental fatigue of eating the same bland meals endlessly. Calorie intake is also very important to keep in mind for emergency food supplies, as keeping your energy up is important. Everything from dried meat to grains to Jell-O can be included in your food store, and probably should be – if it can be stored for a long while, it has a place on your shelves unless you hate it or are allergic to it. Your survival food store should include the following types of food:

- **Grains** – both wheat and corn fall into this category, for flour and corn meal respectively. These can be stored whole and ground, or as flour. Whole grains keep better but are more work to prepare. Included in this category are also various mixes – you can include pancake mix, pie crust mix, and even cake mixes in your grain supply as long as you also have some means of cooking them to the proper temperature.

- **Meats** – a combination of dried meats and tinned meats, as well as canned fish, for the essential proteins that the carnivore known as man requires. Tuna and sardines will last for a minimum of three years in their tins, and provide a lot of essential nutrients. Be sure to have a can opener so that you can open them easily. The author has witnessed Russians opening a stout tin can with a knife – it works, but it produces frequent sliced fingers even among those skilled at doing it.

- **Beans and sprouts** – you should have several types of beans, and possibly the material for sprouts as well.

- **Dairy products** – powdered milk and powdered eggs are the only dairy products you will be able to store over the long-term.

- **Sugars** – actual sugar, as well as hard candy, chocolate powder for cocoa or drink mix, and honey, which is an excellent long-lasting food resource but should not be given to small children. White sugar keeps indefinitely, confectioner's sugar and granulated sugar will last two to three years, and brown sugar lasts around 18 months.

- **Fruits and vegetables** – either dried or canned, these are essential for proper nutrition. Shelled nuts only last a few months before going stale, but those left unshelled will stay edible for at least two years.

- **Condiments and miscellaneous foods** – salt, pepper, onion powder, powdered or dried tomatoes, and whatever spices you enjoy, preferably in quantity and variety. Jell-O and other gelatin or pudding mixes can keep for a year or two. Teabags will keep for around three years, so you can sip a placid, civilized cup of Lipton's or Earl Grey even while the world burns around you. Ritz crackers and Peanut Butter are a basic long-term storage staple.

MRES – MADE READY TO EAT

History of the MRE: *The Meal, Ready-to-Eat is also known as an MRE. It is self-contained in light weight packaging and used in the United States military. The MRE replaced the Meal, Combat, Individual (MCI) rations. Because of the range of geographic tastes, the Department of Defense began to design MREs that would suit many different cultures.*

The goal of an emergency food supply is to have enough calories to survive until other food sources arrive so it is important that the service member eat the entire meal. This did not always happen if the meal did not seem palpable to the person consuming it. Vegetarian options have now been added.

There are no laws that forbid the resale of MREs. They can be found on survival and disaster websites and were used in for the victims in Hurricane Katrina and other disasters where there was a shortage of food.

MRE MEALS
Contents may include:

- Main course or entree

- A side dish

- Dessert or snack

- Crackers or bread

- Spread of cheese, peanut butter or jelly

- Powdered beverage mix: fruit flavored drink, cocoa, instant coffee or tea, sport drink or dairy shake.

- Utensils (usually just a plastic spoon)

- Flameless Ration Heater (FRH)

- Beverage mixing bag

- Accessory pack:

  - Xylitol chewing gum

  - Water-resistant matches

  - Napkins and toilet paper

  - Moist towelette

  - Seasonings include salt, pepper, sugar, creamer and Tabasco sauce

Many items are fortified with nutrients. In addition, the Department of Defense policy requires units to augment MREs with fresh food and A-Rations whenever feasible, especially in training environments.

NUTRITIONAL REQUIREMENTS
- Each meal needs to provide 1,200 calories

- Each has a shelf life of 3 years depending on storage

- It is not recommended that MREs be consumed for more than 21 days in a row

WATER STORAGE AND PURIFICATION
When planning your water resources for survival you need to deal with storing water, finding water and purifying water.

STORING WATER

For your in-home cache or survival retreat stash, you should count on two gallons of water per-person per-day. While this is more water than necessary to survive (except in hot climates or after strenuous exertion) it ensures water is available for hygiene and cooking as well as drinking.

Our personal in-home stash has enough water for a week.

Commercial gallon bottles of filtered/purified spring water often carry expiration dates two years after the bottling date. A good rotation program is necessary to ensure your supply of water remains fresh and drinkable.

If you have a spare refrigerator in the basement or the garage, use PET water bottles (the kind soda or liters of water come in) to fill any available freezer space. In addition to providing you with fresh, easily transportable drinking water, the ice can be used to cool food in the refrigerator in the event of a power failure. We have found that these bottles, which are clear and have screw-on caps like soda bottles, will withstand many freeze-thaw cycles without bursting or leaking. (The bottom may distort when frozen, but this isn't a big problem.) For self-storage of large amounts of water, you're probably better off with containers of at least 5 gallons. Food-grade plastic storage containers are available commercially in sizes from five gallons to 250 or more. Containers with handles and spouts are usually five to seven gallons, which will weigh between 40 and 56 pounds. Get too far beyond that and you'll have great difficulty moving a full tank.

Solutions designed to be added to water to prepare it for long-term storage are commercially available. Bleach can also be used to treat tap water from municipal sources. Added at a rate of about 1 teaspoon per 10 gallons, bleach can ensure the water will remain drinkable.

Once you're in a survival situation where there is a limited amount of water, conservation is an important consideration. While drinking water is critical, water is also necessary for rehydrating and cooking dried foods. Water from boiling pasta, cooking vegetables and similar sources can and should be retained and drunk, after it has cooled. Canned vegetables also contain liquid that can be consumed.

To preserve water, save water from washing your hands, clothes and dishes to flush toilets.

SHORT TERM STORAGE

People who have electric pumps drawing water from their well have learned the lesson of filling up all available pots and pans when a thunderstorm is brewing. What would you do if you knew your water supply would be disrupted in an hour?

Here are a few options in addition to filling the pots and pans:

- The simplest option is to put two or three heavy-duty plastic trash bags (avoid those with post-consumer recycled content) inside each other. Then fill the inner bag with water. You can even use the trash can to give structure to the bag. (A good argument for keeping your trash can fairly clean!)

- Fill your bath tub almost to the top. While you probably won't want to drink this water, it can be used to flush toilets, and wash your hands.

If you are at home, a fair amount of water will be stored in your water pipes and related system.

To get access to this water, first close the valve to the outside as soon as possible. This will prevent the water from running out as pressure to the entire system drops and prevent contaminated water from entering your house.

Then open a faucet on the top floor. This will let air into the system so a vacuum doesn't hold the water in. Next, you can open a faucet in the basement. Gravity should allow the water in your pipes to run out the open faucet. You can repeat this procedure for both hot and cold systems.

Your hot water heater will also have plenty of water inside it. You can access this water from the valve on the bottom. Again, you may need to open a faucet somewhere else in the house to ensure a smooth flow of water. Sediment often collects in the bottom of a hot water heater. While a good maintenance program can prevent this, it should not be dangerous. Simply allow any stirred up dirt to again drift to the bottom.

FINDING OR OBTAINING WATER

There are certain climates and geographic locations where finding water will either be extremely easy or nearly impossible. You'll have to take your location into account when you read the following. Best suggestion: Buy a guide book tailored for your location, be it desert, jungle, arctic or temperate.

Wherever you live, your best bet for finding a source of water is to scout out suitable locations and stock up necessary equipment before an emergency befalls you. With proper preparedness, you should know not only the location of the nearest streams, springs or other water source but specific locations where it would be easy to fill a container and the safest way to get it home.

Preparedness also means having at hand an easily installable system for collecting rain water. This can range from large tarps or sheets of plastic to a system for collecting water run off from your roof or gutters.

Once you have identified a source of water, you need to have bottles or other containers ready to transport it or store it.

PURIFICATION

And while you may think any water will do in a pinch, water that is not purified may make you sick, possibly even killing you. In a survival situation, with little or no medical attention available, you need to remain as healthy as possible.

Boiling water is the best method for purifying running water you gather from natural sources. It doesn't require any chemicals, or expensive equipment – all you need is a large pot and a good fire or similar heat source. Plus, a rolling boil for 20 or 30 minutes should kill common bacteria such as guardia and cryptosporidium. One should consider that boiling water will not remove foreign contaminants such as radiation or heavy metals.

Outside of boiling, commercial purification/filter devices made by companies such as PUR or Katadyn are the best choices. They range in size from small pump filters designed for backpackers to large filters designed for entire camps. Probably the best filtering devices for survival retreats are the model where you pour water into the top and allow it to slowly seep through the media into a reservoir on the bottom. No pumping is required.

On the down side, most such filtering devices are expensive and have a limited capacity. Filters are good for anywhere from 200 liters to thousands of gallons, depending on the filter size and mechanism. Some filters used fiberglass and activated charcoal. Others use impregnated resin or even ceramic elements.

Chemical additives are another, often less suitable option. The water purification pills sold to hikers and campers have a limited shelf life, especially once the bottle has been opened.

Pour-though filtering systems can be made in an emergency. Here's one example that will remove many contaminants:

1. Take a five or seven gallon pail (a 55-gallon drum can also be used for a larger scale system) and drill or punch a series of small holes on the bottom.

2. Place several layers of cloth on the bottom of the bucket, this can be anything from denim to an old table cloth.

3. Add a thick layer of sand (preferred) or loose dirt. This will be the main filtering element, so you should add at least half of the pail's depth.

4. Add another few layers of cloth, weighted down with a few larger rocks.

5. Your home-made filter should be several inches below the top of the bucket.

6. Place another bucket or other collection device under the holes you punched on the bottom.

7. Pour collected or gathered water into the top of your new filter system. As gravity works its magic, the water will filter through the media and drip out the bottom, into your collection device. If the water is cloudy or full of sediment, simply let it drop to the bottom and draw the cleaner water off the top of your collection device with a straw or tube.

While this system may not be the best purification method, it has been successfully used in the past. For rain water or water gathered from what appear to be relatively clean sources of running water, the system should work fine.

Information courtesy of:

http://www.captaindaves.com

PREPARE SPIRITUALLY

The Bible says there is only one way to Heaven.

Jesus said: "I am the way, the truth, and the life: no man cometh unto the Father but by me." (John 14:6)

Good works cannot save you.

"For by grace are ye saved through faith; and that not of yourselves: it is the gift of God: Not of works, lest any man should boast." (Ephesians 2:8-9)

Trust Jesus Christ today! Here's what you must do:

Admit you are a sinner.

"For all have sinned, and come short of the glory of God;" (Romans 3:23)

"Wherefore, as by one man sin entered into the world, and death by sin; and so death passed upon all men, for that all have sinned:" (Romans 5:12)

"If we say that we have not sinned, we make him a liar, and His word is not in us." (1 John 1:10)

Be willing to turn from sin (repent).

Jesus said: "I tell you, Nay: but, except ye repent, ye shall all likewise perish." (Luke 13:5)

Believe that Jesus Christ died for you, was buried, and rose from the dead.

"For God so loved the world, that he gave His only begotten Son, that whosoever believeth in him should not perish, but have everlasting life." (John 3:16)

"But God commendeth His love toward us, in that, while we were yet sinners. Christ died for us." (Romans 5:8)

"That if thou shalt confess with thy mouth the Lord Jesus, and shalt believe in thine heart that God hath raised him from the dead, thou shalt be saved." (Romans 10:9)

Through prayer, invite Jesus into your life to become your personal Saviour.

"For with the heart man believeth unto righteousness; and with the mouth confession is made unto salvation." (Romans 10:10)

"For whosoever shall call upon the name of the Lord shall be saved." (Romans 10:13)

What to pray:

Dear God, I am a sinner and need forgiveness. I believe that Jesus Christ shed His precious blood and died for my sin. I am willing to turn from sin. I now invite Christ to come into my heart and life as my personal Saviour.

THE END GAME

Remember the vision where it was revealed that in October of 2017 (I know the day he was told but I'd rather not mention it at this point, but it is in October and in the very latter part) that Russia, China and North Korea would use a combination of drone submarines and unmarked container ships to engage in a nuclear attack on the US coast. The attack will occur between midnight and dawn, and they will attack both Britain and the U.S. simultaneously. God told him that as soon as the missiles appeared on the horizon, the rapture of the church would occur. The Russians are planning this at a compartmentalized, high and secret level, and the reason for the date they have chosen is that it commemorates the 100th anniversary of the Russian Revolution. Afterwards, with Israel's protectors neutralized, they will invade the Middle East as Israel has a warm water port and the largest oil and gas fields on the planet (now known as the Leviathan gas fields).

What do you think gives these countries the go-ahead? The U.S. has already been knocked out by an EMP blast from the sun itself as a result of the approach of Planet X. Communications are down. Martial law has been declared. Food and water distribution are minimal if they exist at all.

David Daughtrey had the 1996 experience where he was told (21 years before the fact) what would transpire when Planet X arrives in the proximity of the earth – the asteroids colliding with the sun. This is what causes the EMP effect and sends us back to the nineteenth century. Notice he said that "several weeks later" the rapture happens.

This paper is an original work – you've never heard anything like this in church. The Book of Revelation is a cryptogram set by the Almighty, and it is solvable, according to the Book of Daniel (at the very end time). But it's only solvable with much effort, diligence and study.

As of this writing the Red Dragon (Planet X) is within our solar system and approaching Planet Earth.

The next piece of the puzzle is this:

"A great sign appeared in heaven: a woman clothed with the sun, with the moon under her feet and a crown of twelve stars on her head. She was pregnant and cried out in pain as she was about to give birth." Rev. 12:1-2.

The great sign of The Woman as described in Revelation 12:1-2 forms and lasts for a few hours. According to computer generated astronomical models, this sign has never before occurred in human history. It will occur once on September 23, 2017. It will never occur again. When it occurs, it places the Earth immediately before the time of the Sixth Seal of Revelation.

The next event that follows is that Planet X will fully eclipse the Sun and cover the whole earth and full moon in shadow on Thursday, October 5, 2017 (a full moon date). Our clue that this is the date is from Revelation, Chapter 6, verse 12:

I watched as he opened the sixth seal. There was a great earthquake. The sun turned black like sackcloth made of goat hair, <u>the whole moon</u> turned blood red... Rev. 6:12.

October 5, 2017 is the *only full moon* that occurs after September 23, 2017 (the sign of The Woman), but before October 11, 2017, the end of The Woman's pregnancy. Jupiter, known to the Jews as the planet of the Messiah, exits Virgo (the Virgin) on October 11, 2017 after having spent 301 days (December 14, 2016 to October 11, 2017) in the "womb" region of that constellation.

The sign of the Red Dragon appears in heaven <u>after</u> the sign of "the woman." Planet X is visible during this total solar eclipse, as the following verse indicates:

Then another sign appeared in heaven: an enormous red dragon with seven heads and ten horns and seven crowns on its heads. Rev. 12:3.

The Red Dragon will collide with the earth on Thursday, October 5, 2017. The Red Dragon will enter Earth's orbit, and fully eclipse the Sun and cover the entire earth and full moon in shadow on Thursday, October 5, 2017.

You have until October 5th, 2017 to prepare for Planet X.

END NOTES

The articles below, detailing a search for a Planet X, or the 10th planet in our solar system, are speaking of the same planet Sitchin calls the 12th Planet. In his book, *The 12th Planet*, Sitchin explains that the ancient Sumerians counted the Sun and the Earth's moon as planets, and thus the Sun, Earth, Moon, Mercury, Venus, Mars, Jupiter, Saturn, Uranus, Neptune, and Pluto added up to 11 planets. Modern astronomy excludes the Sun and the Earth's Moon, counting only 9 planets in our known solar system.

Astronomy
Search for the Tenth Planet
Dec 1981

Astronomers are readying telescopes to probe the outer reaches of our solar system for an elusive planet much larger than Earth. Its existence would explain a 160-year-old mystery. The pull exerted by its gravity would account for a wobble in Uranus' orbit that was first detected in 1821 by a French astronomer, Alexis Bouvard. Beyond Pluto, in the cold, dark regions of space, may lie an undiscovered tenth planet two to five times the size of the Earth. Astronomers at the U.S. Naval Observatory (USNO) are using a powerful computer to identify the best target zones, and a telescopic search will follow soon after. Van Flandern thinks the tenth planet may have between two and five Earth masses and lie 50 to 100 astronomical units from the Sun. (An astronomical unit is the mean distance between Earth and the Sun.) His team also presumes that, like Pluto's, the plane of the undiscovered body's orbit is tilted with respect to that of most other planets, and that its path around the Sun is highly elliptical.

New York Times
June 19, 1982

A pair of American spacecraft may help scientists detect what could be a 10th planet or a giant object billions of miles away, the national Aeronautics and Space Administration said Thursday. Scientists at the space agency's Ames Research Center said the two spacecraft, Pioneer 10 and 11, which are already farther into space than any other man-made object, might add to knowledge of a mysterious object believed to be beyond the solar system's outermost known planets. The space agency said that persistent irregularities in the orbits of Uranus and Neptune "suggest some kind of mystery object is really there" with its distance depending on what it is. If the mystery object is a new planet, it may lie five billion miles beyond the outer orbital ring of known planets, the space agency said. If it is a dark star type of object, it may be 50 billion miles beyond the known planets; if it is a black hole, 100 billion miles. A black hole is a hypothetical body in space, believed to be a collapsed star so condensed that neither light nor matter can escape from its gravitational field.

Newsweek
Does the Sun Have a Dark Companion?
June 28 1982

When scientists noticed that Uranus wasn't following its predicted orbit for example, they didn't question their theories. Instead they blamed the anomalies on an as yet unseen planet and, sure enough, Neptune was discovered in 1846. Now astronomers are using the same strategy to explain quirks in the orbits of Uranus and Neptune. According to John Anderson of the Jet Propulsion Laboratory in Pasadena, Calif., this odd behavior suggests that the sun has an unseen companion, a dark star gravitationally bound to it but billions of miles away. Other scientists suggest that the most likely cause of the orbital snags is a tenth planet 4 to 7 billion miles beyond Neptune. A companion star would tug the outer planets, not just Uranus and Neptune, says Thomas Van Flandern of the U.S Naval Observatory. And where he

admits a tenth planet is possible, but argues that it would have to be so big –
a least the size of Uranus – that it should have been discovered by now. To
resolve the question, NASA is staying tuned to Pioneer 10 and 11, the
planetary probes that are flying through the dim reaches of the solar system
on opposite sides of the sun.

Astronomy
Searching for a 10th Planet
Oct 1982

The hunt for new worlds hasn't ended. Both Uranus and Neptune follow
irregular paths that observers can explain only by assuming the presence of
an unknown body whose gravity tugs at the two planets. Astronomers
originally though Pluto might be the body perturbing its neighbors, but the
combined mass of Pluto and its moon, Charon, is too small for such a role.
While astronomers believe that something is out there, they aren't sure what
it is. Three possibilities stand out: First, the object could be a planet – but
any world large and close enough to affect the orbits of Uranus and Neptune
should already have been spotted. Searchers might have missed the planet,
though, if it's unusually dark or has an odd orbit.

NASA has been recording velocities for a year now and will continue for as
long as necessary. This past spring, it appeared that budget cuts might force
the end of the Pioneer project. The space agency now believes that it will
have the money to continue mission operations. Next year, the JPL group
will begin analyzing the data. By the time the Pioneer experiment shows
results, an Earth-orbiting infrared telescope may have discovered the body.
Together, IRAS and the Pioneers will allow astronomers to mount a
comprehensive search for new solar system members. The two deep space
probes should detect bodies near enough to disturb their trajectories and the
orbits or Uranus and Neptune. IRAS should detect any large body in or near
the solar system. Within the next year or two, astronomers may discover not
one new world, but several.

New York Times
January 30, 1983

Something out there beyond the farthest reaches of the known solar system seems to be tugging at Uranus and Neptune. Some gravitational force keeps perturbing the two giant planets, causing irregularities in their orbits. The force suggests a presence far away and unseen, a large object that may be the long-sought Planet X. The last time a serious search of the skies was made it led to the discovery in 1930 of Pluto, the ninth planet. But the story begins more than a century before that, after the discovery of Uranus in 1781 by the English astronomer and musician William Herschel. Until then, the planetary system seemed to end with Saturn.

As astronomers observed Uranus, noting irregularities in its orbital path, many speculated that they were witnessing the gravitational pull of an unknown planet. So began the first planetary search based on astronomer's predictions, which ended in the 1840's with the discovery of Neptune almost simultaneously by English, French, and German astronomers. But Neptune was not massive enough to account entirely for the orbital behavior of Uranus. Indeed, Neptune itself seemed to be affected by a still more remote planet. In the last 19th century, two American astronomers, William H. Pickering and Percival Lowell, predicted the size and approximate location of the trans-Neptunian body, which Lowell called Planet X. Years later, Pluto was detected by Clyde W. Tombaugh working at Lowell Observatory in Arizona. Several astronomers, however, suspected it might not be the Planet X of prediction. Subsequent observation proved them right. Pluto was too small to change the orbits of Uranus and Neptune, the combined mass of Pluto and its recently discovered satellite, Charon, is only 1/5 that of Earth's moon.

Recent calculations by the United States Naval Observatory have confirmed the orbital perturbation exhibited by Uranus and Neptune, which Dr. Thomas C Van Flandern, an astronomer at the observatory, says could be explained by "a single undiscovered planet". He and a colleague, Dr. Richard Harrington, calculate that the 10th planet should be two to five times more massive than Earth and have a highly elliptical orbit that takes it some 5 billion miles beyond that of Pluto – hardly next-door but still within the gravitational influence of the Sun.

US News World Report
Planet X – Is It Really Out There?
Sept 10, 1984

Shrouded from the sun's light, mysteriously tugging at the orbits of Uranus and Neptune, is an unseen force that astronomers suspect may be Planet X – a 10th resident of the Earth's celestial neighborhood. Last year, the infrared astronomical satellite (IRAS), circling in a polar orbit 560 miles from the Earth, detected heat from an object about 50 billion miles away that is now the subject of intense speculation. "All I can say is that we don't know what it is yet," says Gerry Neugesbeuer, director of the Palomar Observatory for the California Institute of Technology. Scientists are hopeful that the one-way journeys of the Pioneer 10 and 11 space probes may help to locate the nameless body.

The Mysterious Death of Dr. Richard Harrington by John DiNardo
Janice Manning

Source

http://yowusa.com/planetx/2008/planetx-2008-05b/1.shtml

BIBLIOGRAPHY

Brussell Sprout. (2016, January 3). Gigantic Wrecking Ball Headed Toward Earth. Retrieved from http://brussellsprout.blogspot.com/2013/11/gigantic-wrecking-ball-headed-toward.html

0921

Made in the USA
Middletown, DE
22 February 2017